Industrial Exhaust Hood & Fan Piping

A treatise on the Planning and Installing of Fan-Piping in All its branches

W. H. Hayes

Second Edition, Enlarged

Wexford College Press
2003

PREFACE TO FIRST EDITION.

The author of this work is possessed of rare qualifications to treat his subject in a thoroughly practical manner, through long and diversified experience acquired during his connection with one of the largest blower concerns in the United States. Upon the urgent request of the publishers he was induced to prepare the treatment, here comprehended in chapters one to thirteen, for presentation in the columns of SHEET METAL, then known as THE SHEET METAL SHOP, from which source the articles are now compiled and republished, together with a number of correspondents' letters and the author's answers thereto, also valuable data on installing and estimating the cost of an exhaust system by "M. F. H.," an interested and well informed reader who volunteered this additional information from his own intimate acquaintance with the subject.

The increasing utility of blower and exhaust systems for purposes of heating or for carrying away the waste products of machinery employed for manufacturing purposes, etc., brings to sheet metal operators a growing field of opportunity which is, however, accompanied with exceptional responsibility incident to properly designing and

installing this class of work.' In view of the scarcity of literature on this subject and the difficulty
heretofore of obtaining comprehensive working
data it is believed that a large body of sheet
metal workers will find the book a welcome aid
and it is hoped that many who have been deterred
from taking up blower and exhaust work will,
through the assistance of this manual, be encouraged to devote attention to this very important
branch of the business.

THE PUBLISHERS.

PREFACE TO SECOND EDITION.

During the interval of reprinting, made necessary to meet the continuous demand for this book,
we have taken advantage of the opportunity to
introduce in the present edition much interesting
discussion of problems encountered by correspondents, continued subsequently to the publication of the first edition; also a practical article
on the Details of Settling Chambers for Grinders,
covering the methods of construction. The additional text makes the work considerably more
valuable as a repository of information on the
subject.

THE PUBLISHERS.

CONTENTS.

APPENDIX.

CHAPTER I.

GENERAL RULES.

The aim of this work is to treat the subject of fan-piping in all its branches, from a strictly practical point of view. It is assumed that the reader who is to derive benefit from this discourse has a general knowledge of sheet metal work as applied to the laying out of pipes, elbows, etc. It is not the purpose to submit any scientific data pertaining to the number of cubic feet of air any certain style of blower will handle in a given time. Any blower company will furnish such data on application.

It is with the proper method of planning and installing exhaust systems that we shall deal.

The matter of general rules should be first considered, rules applicable to any and all cases, whether the plant is to handle shavings, emery dust, cement, wood pulp, or any one of the numerous materials that blowers are now required to handle. These rules apply to general construction.

Always set the blower as near the heaviest work

as it is possible to get it; and be sure that the shaft at that place is heavy enough to drive it.

Never unite two or more discharge pipes into one, but run them all independently to their destination. This does not apply to the double-exhauster, where the two outlets may be connected in the form of a Y-branch, since both fans are run by the same pulley. In the case of several fans in different parts of a plant, running independently, each having the capacity of a 14-inch pipe, let us suppose that two out of three fans should be out of commission and that the three pipes branched into one main whose capacity would of course be equal to the combined area of the three branches, that is to say 24 inches. In such case the remaining blower would soon be in trouble and there would be plenty of trouble in store for the other two; for the large main pipe would be packed full of material, and it would be difficult to clean it out.

Never use square pipe unless its use is an absolute necessity as may sometimes be the case in getting by a close place. Round pipe is cheaper, neater in appearance and more efficient since there is much less friction created in it, and friction is to be avoided where it is possible to do so.

Never flatten a pipe to get between joists, or for any other reason. To flatten it in the least is to reduce its area and consequently to defeat the object sought in proportioning it. Transform it if necessary to a rectangle or a oblong, preferably the latter. Make the transformer a little in excess of the area of the pipe and get back to the original pipe as soon as possible, avoiding abrupt angles in doing so.

Never use a short radius elbow where its use can be avoided; an elbow will cause more friction than will ten times its length of straight pipe.

Never use a so-called "peened" elbow, by which I mean one with locked seams, if the object is to erect solid, substantial work.

If the reader has had any experience at this work, it has probably taught him, to his cost, that when a machine operator discovers that his pipe has become clogged, he shuts down his machine and goes at it with a club, and the elbow generally gets the first crack. "Peened" elbows generally disintegrate under this treatment. There is a popular saying that "the best is the cheapest," which, while it may not be universally true, certainly holds good here.

In an attempt to erect a cheap job, by which is meant soldered joints, peened elbows tied up with wire, slip joints, etc., it will be discovered that the job is only begun by the time it should be finished, that is if the plant is in operation while erecting the work, as is generally the case.

With the pulling and yanking generally indulged in during the act of raising and lowering hoods, and the throwing off of belts that may fall across the pipe, it will be damaged almost as fast as it can be repaired, unless it is as rigid as it can possibly be made.

Too much stress cannot be laid upon this fact. It has been the writer's experience as one of some forty or more erecting men "on the road" for the same concern, that those who kept it always in mind to put everything up to stay, not only "brought home the bacon" but got the job finished quicker. Otherwise it would have been necessary to go back and overhaul the first pipes, by the time the last were in place.

There is no light sheet metal work known that is subject to more rough usage than a piping system, especially in a large planing mill.

Not only is the material it is handling rapidly destructive of the pipes through impact on the curves and angles, but, as before intimated, the operators are never very tender in their treatment of them.

Never use "grooved" pipe; rivet all work substantially and solder all joints where practical.

It will be suggested that some joints cannot be riveted. In such cases they may be bolted through the pipe as will be explained in the proper place.

All elbows should be made of material at least two gauges heavier than the pipe.

All main pipes should be riveted throughout and fitted with hand holes about 15 feet apart for cleaning out. These hand holes should be covered with slides. It is good practice to rivet the sections of main pipe together by means of these hand holes, holding on the rivets with a "dolly" bar, since at the end farthest from the fan the pipe will be too small for a helper to enter.

All tee branches should enter the main pipe at the side or top, never under any circumstances from the bottom. These should always be riveted on and never notched in, as the blocks and chips would soon play havoc with the notches. Tee branches should always point at an angle of at least 45 degrees toward the exhauster.

Never connect a suction pipe directly to the exhauster. What is known as a "fan joint," which is really a collar, clamped at the seam and easily removable, should always be used. When this is removed there should be a space of at least 12 inches between the end of the suction pipe and the inlet of the exhauster.

Thus, if the exhauster is out of commission, a set screw loose, a few rags riding the hub or a blade broken, repairs may easily be made without taking down the pipe.

All hoods and hoppers, no matter what sort of material they are handling, must be assured of an air space at least equal in area to their outlet. Thus, it is not good practice to fit the top hood of a planer, sizer, or matcher too snugly around the pressure bar. If, for example, a 7-inch pipe is attached to the hood, there must be space around the bottom of that hood sufficient to let in enough air to keep the material moving through the pipe.

"Keep everything moving," should be the slogan of the blow-pipe-man; although to succeed in doing so is no light matter.

Whatever may be handled, let the aim always be to give it the most direct route to its destination. Put no corners, abrupt angles or pockets in its way and, above all, let in the air. To let out the air is no less important. I have seen systems in the South and West before the dust separator had made its advent in those regions, proportioned well enough and fairly erected, so far as the interior of the plant was concerned, but rendered practically worthless by the manner of depositing the material.

A 20-inch discharge pipe would enter a vault 20 feet high and 10 feet square and, on the other side, higher up, or perhaps lower down, would be found a round hole also 20 inches in diameter.

The purpose of the hole was to let out the air— to make a vent; a good idea as far as it went, but the hole should have been at least three times as large.

The air, when starting, was forced into the vault at a high velocity; perhaps there was a pressure of five ounces to the square inch behind it. Once inside the big enclosure, of course, it expanded and lost its velocity. As soon as the vault filled with air it became necessary for the exhauster to force it out of the other hole at the same velocity. This was manifestly impossible.

The result was that the air not being able to get

out fast enough would not come in fast enough and, of course, the suction was correspondingly weak.

While as much power was required to run the exhauster as would be necessary under favorable conditions, it was not working up to half its maximum capacity.

I have given this illustration to show that it is just as important to let out the air as it is to let it in.

CHAPTER II.

The most important step after locating the exhauster is the dust separator, this is also practically the starting point of the job.

In Fig. 1 is shown a dust separator as installed over the boiler house of a wood working factory, in which A is the transformer connecting the round discharge pipe from the exhauster to the inlet H. Incidentally, these transformers should always be connected, as shown in Figs. 1 and 2, that is to say, straight on the top, and the outside line of transformer should be tangent to the shell of the separator as shown at B in Fig. 2.

The elbow C in Fig. 2 shows the wrong way to connect a separator. As indicated by the arrows, the material follows the outside curve of the elbow and the lightest of it at least will thus be caught in the inside current and blown out of the air outlet before they have had a chance to strike the outside shell D. The dotted lines show the proper way to connect to the separator, and the dotted arrows show the material traveling as it should.

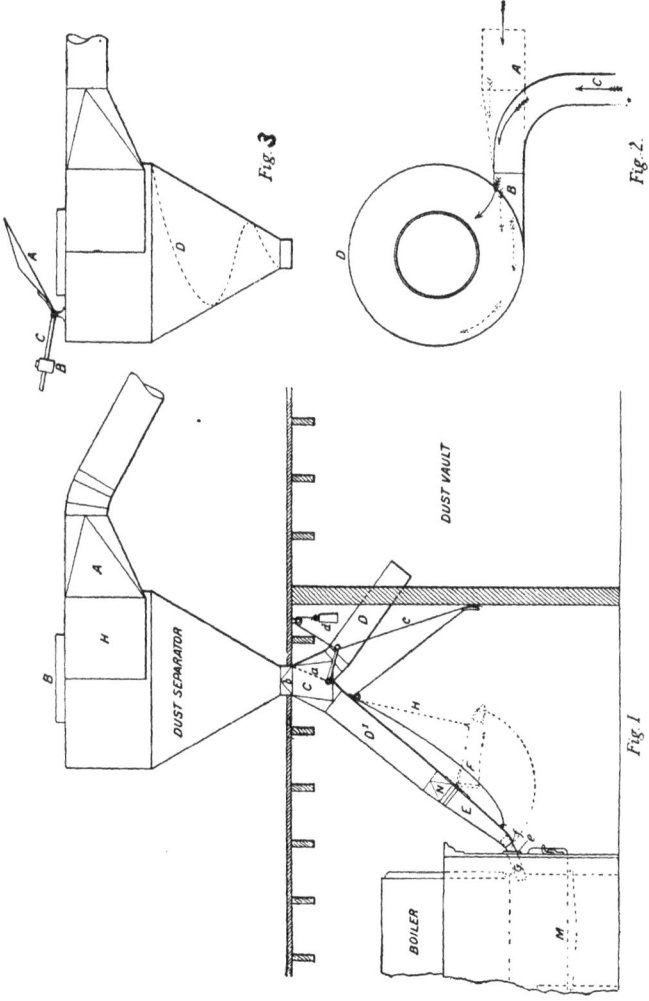

Fig. 3

Fig. 2.

Fig. 1

DUST VAULT

DUST SEPARATOR

BOILER

DETAILS OF DUST SEPARATOR, SWITCH AND FEEDER.

After the transformer is put on, the pipe may be connected at any angle.

A single furnace feed is shown in Fig. 1, in which the material is supposed to drop quietly into the furnace and indeed it is quite necessary that it should do so since it is bad practice for several reasons to blow air over the top of the fire. The location of the dust separator should never be made a matter of guesswork, or haphazard calculation, for upon its location depends the successful working of the feeder. The main point is to get fall enough to drop the material from the separator to the grate bars at M under the boiler. In such a case as is here shown, the problem is not so difficult, but it becomes tremendously so when we have several boilers side by side, all to be automatically fired with but one separator to do the work. The principle involved, however, is shown in the illustration, and the question of firing more than one furnace will be taken up later.

At the bottom of the dust separator is shown a switch C. Let it be remembered that the pipes D and D^1 are round, as is also the bottom of the separator, so the switch must be a three way, flat on the sides and square at *b,* to insure the successful operation of the gate, which is indicated by the dotted line *a.*

The gate is operated by a handle set at right angles to it as shown, which is pulled down by the line *c* and up by the weight *d,* which is made operative by fastening it to a rope which is passed through a pulley or shieve attached to the ceiling.

This switch should be made of at least sixteen gauge iron and only level sheets should be used, as it is necessary that the gate should work easily and yet be practically tight. It will be seen that by operating this gate the material may be deposited either into the furnace or the dust vault.

The pipe entering the dust vault is shown at D, while D^1 is the pipe that enters the furnace. It will be seen that this pipe is transformed again at N from a round to a square, and the square is then gradually changed to a rectangle upon entering the furnace. The cast iron chute G is bricked into the front wall of the furnace, passing through the cast iron front and is held in place by a flange *e.* The size of this chute, and consequently the end of the feeder, depends entirely upon the amount of material to be handled, since there is no air going through it. The average feeder would measure about 10 inches by 5 inches, the 10-inch side of course being parallel with the grate bars.

The material outlet *b* at the bottom of the separator is 12 inches in diameter. The two pipes D and D^1 are also each 12 inches in diameter, considerably larger, indeed, than the 5 by 10-inch hole. The reason for this is that the material as it is leaving the bottom of the separator, is whirling around rapidly as it falls and would pack and clog in a smaller pipe. As it gets further down it loses its momentum and falls by gravity. At the point where the feeder enters the chute is shown a telescope joint *f*; this should be made to fit the chute snugly and slide up into the feeder far enough to let it fall out of the hole. It will be seen that the feeder E and transformer N are connected by hinges. The dotted lines at F show the feeder disconnected from the furnace and swung back out of the way.

If after the system is in operation it is found that the material does not fall back on the grate bars far enough, it will be necessary to get a little air through the feeder pipe. Fig. 3 shows the wrong way to do that. A represents an automatic cover held open by the action of the air escaping from the separator and I am aware that this scheme is practiced even now by experts. The idea is to move the counterbalance B on the bar C until the cover will press on the outgoing air just hard enough to force some of it down through the feeder. The effect of this operation is the same as that referred to in the previous article with reference to blowing the material into a vault where the vent was too small.

The dotted lines at D show a spiral strip that may be riveted around the inside of the cone. This piece need not be more than 3 inches wide and the length of it should be governed by the amount of air needed.

It must start from the bottom, but unless an unusual quantity of air is needed it need not go to the top of the cone. In putting this in, care must be taken to do a smooth job, as stringy material is apt to get caught on a projection and ultimately clog the separator.

If a cover is needed on the top of the separator to keep out the rain and snow a canopy is as good as anything, but it must be placed high enough to allow free egress to the escaping air.

CHAPTER III.

The present chapter shall deal with the automatic firing of three boilers with material delivered through one dust separator. The front view of the layout is seen in Fig. 5, while Fig. 4 shows a side view, that is, a view looking parallel with the boiler fronts. The separator is not shown in the sketch. Neither is the main switch which divides the material as it leaves the separator and deflects it—either into the feeder system or the dust bin, as this switch, which is above the roof, was shown in the foregoing chapter.

It will be seen then that this plant requires four switches: the main switch under the separator and those designated by the letters, A, B, and C, in Fig. 5. The gates of these switches, shown by the dotted lines, *a, b,* and *c,* are shown half and half, that is to say, they stand in the center of the inlet in each case. The arrows show the material leaving the main switch as being divided by the gate *a* in switch A, traveling through the pipes D and E, being again divided by gates *b* and *c* in

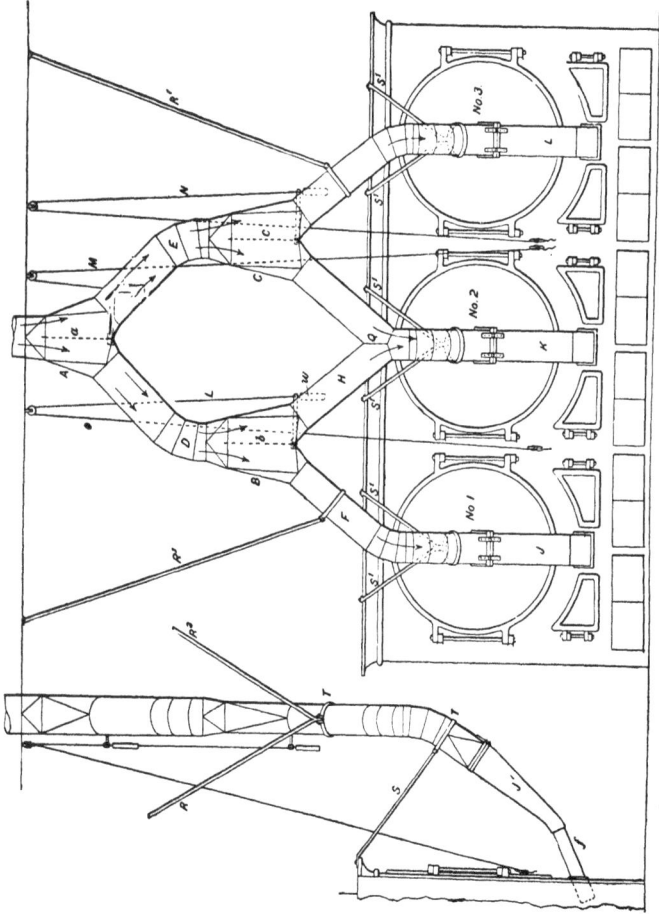

Fig. 4.

Fig. 5.

SIDE AND FRONT VIEW OF SWITCH SYSTEM FOR FIRING THREE BOILERS.

switches B and C, and thus flowing through the feeders proper, J, K, and L. I do not claim that setting these gates at a middle position will divide the material equally. As has been stated, the material on leaving the separator is whirling rapidly. This is caused by the centrifugal motion created in the separator, which motion is the prime factor in the separating process. The material, as it travels through the pipe, assumes a spiral motion and the only way to ascertain when the proper amount is entering each furnace is to watch the fire and move the gates to right or left, until the desired result is obtained.

To explain the use of these three switches, let us suppose that boiler No. 1 is laid up for repairs. We are then only firing boilers Nos. 2 and 3. To do this, throw the gate a, of the top switch A, to the left, by pulling the cord M, and throw gate b in switch B to the right by letting the lever drop; or in other words, by slackening cord L, when the weight w will pull down the lever. By throwing gate a to the left, the material is caused to flow into pipe E and is divided and thrown into boilers 2 and 3, by gate c in switch C.

The reason for closing gate b in switch B is to keep any dust that may leak from the switch above or back up from the Y-branch Q, from drifting through the pipe F and thus getting into the boiler room, since it is likely, if boiler No. 1 is being repaired, that its feeder J will be thrown back and hung out of the way.

Another way to accomplish the same result is to leave gate *a* as it is shown in Fig. 5, and operate gates *b* and *c,* as just described.

Thus the reader will see that any one or two of these boilers may be cut out by manipulating the gates or all may be cut out by closing the gate in the main switch under the separator, thus causing the material to flow into the dust vault.

I have endeavored to show in both Figs. 4 and 5 the manner of bracing this feeder system.

R, R^2 and S in Fig. 1 are braces made of wrought pipe. R and R^2 are supposed to be anchored to the ceiling, or to any convenient joist or girder above the work. In the front view of Fig. 5 the braces are shown by R^1 and S^1.

The size of the pipe used for these braces, of course, depends upon their lengths; and if the reader wonders why such heavy bracing is used at all, the answer is rigidity.

It not only has to support the pipe, switches, feeders and everything below the roof which may, and often does, fill with shavings when the mill is taking heavy cuts and the fireman is caught napping, but it must hold it perfectly rigid. In a case such as we have just referred to, suppose the fireman sees smoke coming from the mouths of his feeders. This means they are clogged, and not only clogged but hot. His first move is to disconnect them. He does this by pulling up the slip point *f* as has been explained, and proceeds to clean them out.

Then he swings them up to the front again and lets down his slip.

This must all be done quickly, for every moment lost means loss of steam, and besides, there is dust flying all about the room.

If there is the least play in that system at any point, those sliding joints will not go back into place without more or less wedging and prying; whereas if the pipes are perfectly solid the slips may be entered as easily as they were taken out.

Of course the manner of bracing this work must depend upon conditions and no set rule can be given, but no fear need be entertained of getting it too strong.

The usual practice is to make draw bands, as shown at T and T of Fig. 4, using ¼ by 1¼-inch bar iron, clamped with ⅜-inch bolts. In making the brace, first get its length, then heat the pipe and flatten at each end, drilling or punching one end for the ⅜-inch bolt and the other end for two ⅜-inch lag screws.

In the present case S and S¹ show the feeder braced to the top of the cast furnace front, which is good practice where the boilers have such a front to utilize. In this case the casting may be drilled and tapped for a ⅝-inch cap screw. The braces should be well spread, as shown in Fig. 5.

As regards the size of the pipes to be used in a feeder system, this matter also is governed by conditions. It may, however, be borne in mind that a 12-inch round pipe will deliver all of the shavings that one furnace will burn, and this requires a feeder 11½ by 11½ inches in size, tapering to 5 by 12 inches. It may sometimes happen that there is not sufficient room directly under the boiler for a feeder nozzle. In that case the pipe must be divided and a hole 4½ by 9½ inches cut out on each side of the center, as near the boiler as necessary and two 10-inch pipes used, the feeders being made 9½ by 9½ inches, tapering to 4 by 9 inches.

It will be seen in Fig. 2 that the pipes D and E are considerably larger than the pipes supplying the feeders J, K and L. This is because either of these pipes may, in an emergency, be used to supply two feeders.

In this case the large pipes should be twice the area of the small ones.

In Figs. 4 and 5 are shown semicircular doors swinging from hinges attached to the fronts. These are really the flue caps. Care must be taken to place the hinged joint of the feeder J^1, Fig. 4, a sufficient distance from the front to allow these doors to swing open.

Lastly, the elbow switches and feeders on this job should be made of No. 16 galvanized iron, the gates still heavier, and the pipe not lighter than No. 20 galvanized. All joints must be securely riveted and soldered and tightly covered hand holes must be placed at convenient distances for cleaning out purposes.

Perhaps the idea will occur to the reader that

one of these switches is superfluous, that the same work would be done with the top switch and switch B, for instance, letting the supply pipe to boiler No. 3 run from the top switch.

It must be borne in mind that the air needed to clear all of these feeders is primarily controlled by the top switch and if a light feed should be wanted in boiler No. 3 and the top switch thrown to the right so far as to almost close it, an excessive quantity of air would be blown under boilers 1 and 2. Whereas, in the present layout, if too much air or too much material is carried either to right or left, it can still be regulated or deflected into boiler No. 2, or taken from it as the case may require.

CHAPTER IV.

In the present chapter we shall try to show in detail the manner of constructing the feeder nozzle and switch previously described; also the long shavings switch shown in Fig. 9.

Mention has been made of the slip joint of the feeder, which is here shown at A and A¹ in Fig. 6. This slip joint is made of material of the same gauge as the feeder proper, that is, No. 16 galvanized iron, riveted smooth inside and stiffened at the upper end by a band of ¼ by 1-inch iron as shown at C and C¹.

This slip is made ⅛ inch larger all around than the feeder B and B¹ to insure easy working and has two catches, D in side elevation and D¹ in the bottom view, bent at the ends as shown. Two clips E and E¹ are shown riveted to the feeder bottom, flush inside, through which these catches slide.

By pulling on these catches by means of the bent ends it will be seen that the nozzle will slide up until the band C strikes the clip E, and this operation will allow the feeder to drop out of the hole.

It will be seen that the space F F¹ on the feeder proper must be made perfectly straight. Since a 3-inch slip is sufficient to hold the feeder in the casting, the space between F and F¹ must be 6½ inches. Then from the line F¹ E of the side ele-

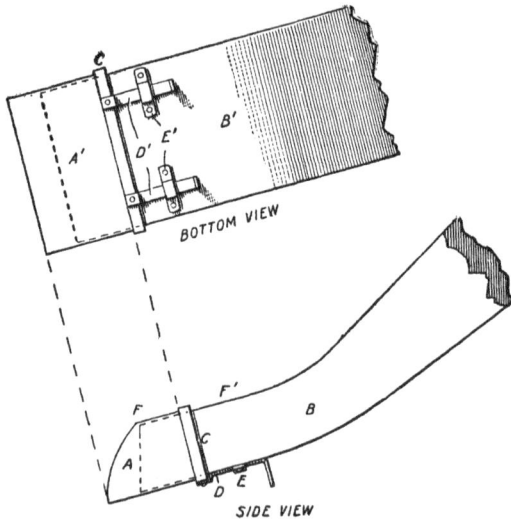

FIG. 6.—DETAIL OF FEEDER NOZZLE, SHOWING SLIP JOINT.

vation the top and bottom curves of the feeder will start at whatever radius they require.

In Fig. 7 is shown the proper method of constructing what is generally known in blow-pipe parlance as the "regular" or short shavings switch.

What we aim to show is the best method of con-

PLAN

VERTICAL SECTION

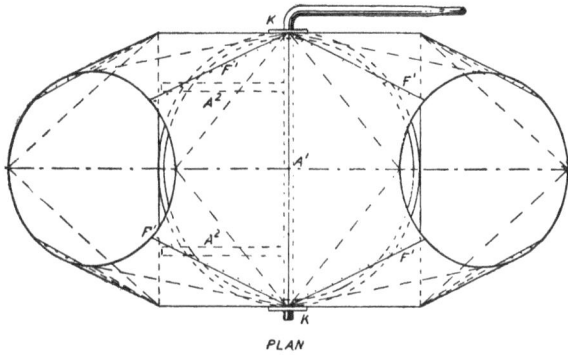

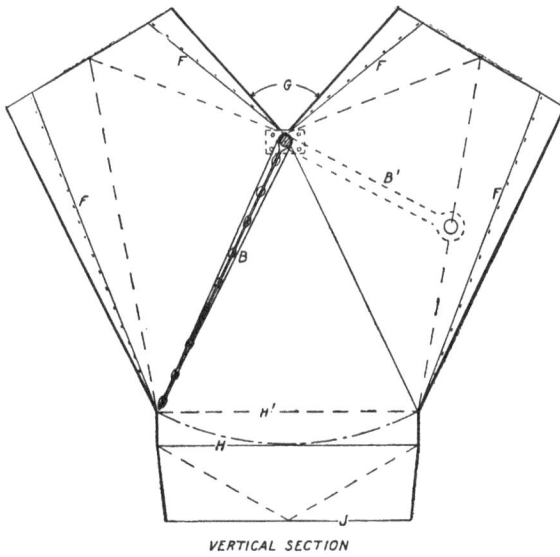

FIG. 7.—DETAILS OF "REGULAR" SWITCH.

structing it. In the upper drawing is shown a plan of the switch, the dotted lines A^1 A^2 A^2 showing the gate, while the lower drawing shows a section on the center line of the plan. In this view the gate is shown at B, thrown to the left, the handle being shown by the dotted lines at B^1.

In Fig. 8 is shown the manner of constructing the frame for this gate. C and C^1 show a ¾-inch round bar bent at right angles, C^1 being the handle with a welded eye C^2 at its end, which should be three-quarters of an inch in diameter. The distance designated D should be the width of the body of the switch.

A sheet of No. 14 gauge iron long enough to reach from the top to the bottom, as shown by D^1, bent around the bar C and brought even with the top again and made ¼ inch narrower than the inside measurement of the switch, constitutes the body of the gate. This may be riveted through the bars E and E and must be closed together and riveted on the sides, as indicated in Fig. 7. It will be seen that by bending this around the bar C the gate on the outside is made perfectly smooth.

In Fig. 7 are shown the riveted lap joints in the body of the switch. These are designated by F in the section and F^1 in plan. Thus it will be seen that this switch, to be made properly as well as economically, must be made in five pieces and have all laps on the corners. The throat G is one piece and this piece must be put in last after the gate B is fitted.

It will readily be seen that there can be no lap
between the points K and K, since the gate must
swing between these points. It will also be seen
that the space between lines H and H[1] is necessary
between the body of the main switch and the inlet
J for the end of the gate to swing in, and in view
of these necessities I think the reason for making
the switch in five pieces becomes apparent.

The bars E and E in Fig. 8 are of ¾ by 1-inch
iron scarfed to nothing on the upper end as shown
and welded to the round bar C.

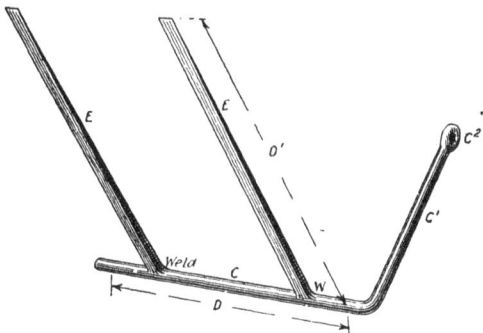

FIG. 8.—FRAME WORK FOR GATE.

In Fig. 9 is shown what is generally termed the
long shavings switch, and is the only switch that
can be used successfully in stave mills, barrel factories, tub factories and mills where dressed cot-
tonwood products are handled, or any plant
where long, stringy shavings are made.

I have shown both of these switches as they
would stand on the shop floor in course of erection
upside down. The inlet or top of the switch is

shown at F, while G and G designate the two outlets. B and B show the pockets directly under
which the top bars of the gate are located.

The handles of these gates, C^1 and C^1, instead
of being at right angles to the gates, as in Fig. 7
and 8, are parallel with them and operated by
having the cross bar H H^1 bolted to them. As
these gates both swing together. one of them when

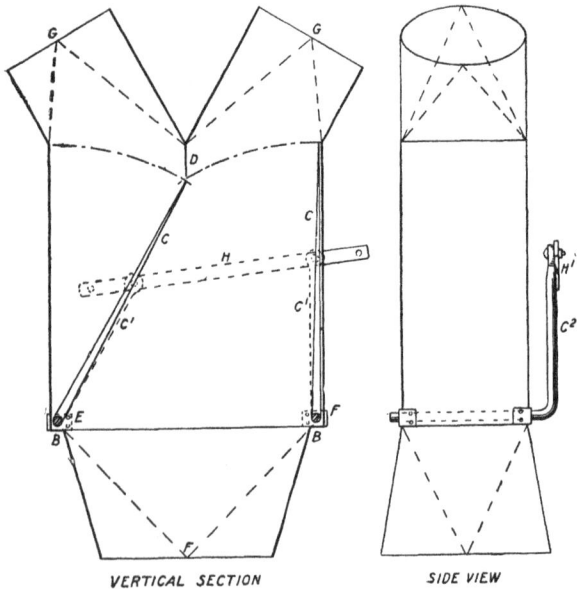

VERTICAL SECTION   SIDE VIEW

FIG. 9.—DETAILS OF LONG SHAVINGS SWITCH.

set will be directly over the point D. It will thus
be seen that there is no chance for any material
to get caught anywhere. The reinforcing pieces
shown at E E should be made of at least ⅛-inch
material, since they are really the bearing for the

gate bars. Reinforcing pieces are also shown in Fig. 7 at K and K.

A division plate is shown at D reaching to within 1 inch of the end of the gate C. This inch of space is necessary, since, when operating the gates, a long shaving or a bunch of them might catch at that point, while with that inch of space they will not be pinched by the gate and will ultimately fall off. These gates must always be thrown completely over and never placed in an intermediate position, since the stringy shavings will catch easily on the division plate D,

CHAPTER V.

In Fig. 10 is shown the plan of a piping system for a planing mill. The object of this sketch is to show the manner of tapering the pipe, the leads from each hood, and the proper location of the exhauster.

As shown in this plan, there are five heavy machines in the mill. First, a "sizer," which is a large, four-headed planing machine for dressing what is called "dimension" timber. Sizers are used largely on this account for dressing ship timbers.

The next is a double surfacer, which is a top and bottom head machine. Then there are two flooring machines, used for getting out all sorts of tongued and grooved stuff. The band re-saw on the end is also a large contributer when ripping timber for the sizer and surfacer. Note that the fan is set as near as possible to the heavy work. Care must be taken also to get a good long drive for the belt, which must be a heavy belt, even if it is necessary to use a counter-shaft. It is probably not generally known that a blower requires an amount

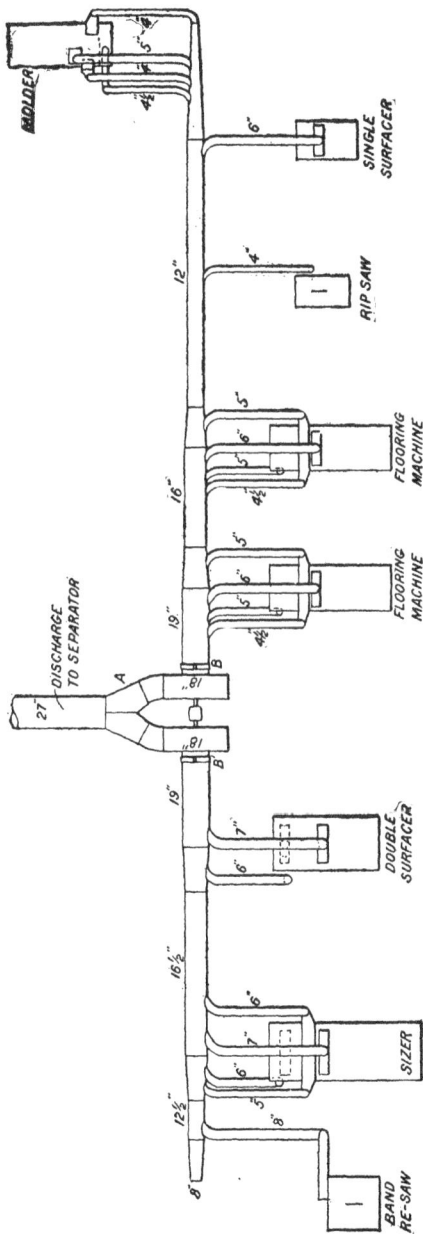

FIG. 10.—PLAN OF PIPING SYSTEM FOR PLANING MILL.

of power out of all proportion to the size of it. The blower shown in this sketch would require no less than 20 horse power if kept up to speed on heavy work.

To determine the size of the blower needed, first determine the size of the pipes. And right here I can offer a bit of advice. First find the men that run these machines and talk to them about them. I learned long ago the wisdom of listening to their suggestions.

For instance, in the present case there would probably be a man about that mill who had run the sizer or one like it. He can tell what class of work it does and how heavy the cuts are. For this machine I have shown a 7-inch top pipe, a 6-inch bottom pipe, 6-inch heavy side pipe, and a 5-inch light side. This is about what the average timber dresser would require, but there are exceptions.

On any four-headed machine, the top head does the heavy cutting, the bottom head the surface cutting. Also, one particular side head knocks off the big chips, while the other simply cleans up. That is why, on the sizer and flooring machines, one side pipe is shown larger than the other.

All flooring machines, or matchers, as they are sometimes called, will require the sizes shown, namely: Top, 6-inch; bottom, 5-inch; heavy side, 5-inch; light, 4½-inch.

A 7-inch pipe will handle the cut from the top head of a double surfacer unless it is used for dressing timber wider than 20-inch, in which case an 8-inch pipe should be used.

A 7-inch pipe, if the hood is properly made and fitted, would do for a light cut up to 24-inch, but in some cases this machine does the timber dressing. 4 by 6-inch timbers are shoved in side by side until the whole width of the knife is taken up.

Now, having determined the size of each pipe, find their areas and add them together. In the present case the combined areas are a little in excess of that of two 18-inch pipes, so we have taken the next highest unit, 19-inch, and used a fan with 18-inch inlets and outlets.

By doing this the maximum of efficiency is obtained with the minimum of power. We have figured from the basis of 19-inch. Starting on from B of Fig. 10, substract the combined areas of a 6-inch, two 5-inch and one 4½-inch pipe from the area of the 19-inch, and the result will be a trifle less than the area of a 16-inch pipe, and so on.

If the answer comes above any figure, take the next higher.

Now as to tapering, never start to diminish the capacity at any point between the branch and the fan. Tap into the straight pipe and reduce beyond it unless, as shown in Fig. 10, there are four branches in close proximity. In this case, of course, one or even two may be tapped into the tapering part; in fact, they should be, but be sure that the capacity of the taper at the point tapped

at least equals the capacity of every pipe beyond it.

In Fig. 11 is shown what is known as the telescope hood. Such hoods are for the side heads of
a flooring machine. B and B¹ show the cutter
head, while the arrow shows the direction of its
rotation. The collar for the pipe shown at D and
D¹ should always be placed as shown, so that the
suction will be strongest at the point where the
chips leave the knife. One of these hoods fits over
the other and each hood will follow the knife out

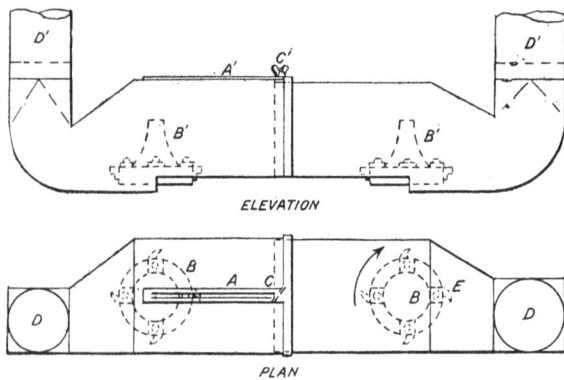

ELEVATION

PLAN

FIG. 11.—DETAILS OF TELESCOPE HOOD.

or in to suit the varying widths of boards, and
hold in position, after everything is set, by tight-
ening the thumb screw C C¹. This screw is fastened securely near the end of the inside hood, and
a reinforced slot as long as desired is put in the
outside one.

In Fig. 12 is shown a design which is a good

one for the top head of any machine whose chip breaker curves over the knives as shown at A.

It will be seen that the pipe B is placed in a line with the flow of material, which will be thrown off at an angle by the curved chip breaker.

At A in Fig. 10 is shown a Y branch, connecting two 18-inch outlets to one 27-inch discharge, the 27-inch discharge having the capacity of two 19-inch pipes. This Y branch should have a narrow spread, so that the material, when it enters on one

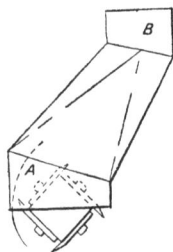

FIG. 12.—HOOD FOR A TOP CUTTER HEAD.

side, will not counteract upon that entering the other. At B and B¹ are shown the "fan joints" spoken of before.

It will be seen that by bracing this work properly, repairs to the fan may be easily made. The discharge is always bolted on with cap screws. These may be taken out, the fan joints taken off, when the fan may be lifted entirely clear of the pipes.

CHAPTER VI.

PIPE CONNECTIONS FOR A FLOORING MACHINE.

In this chapter we shall give attention to the details pertaining to an individual machine. Fig. 13 shows the connections for the top and two side heads of a flooring machine. The pipe from the bottom head is generally run across the floor, or under it, if possible, to a convenient upright, and then up and across to the main. There is no trick in connecting this bottom head, but the construction must be such that the pipe can be easily disconnected in case it becomes filled with shavings. It will be seen that no ''cut offs'' or ''gates'' are shown in these sketches. I do not think they should be used except on floor sweepers or for some machine for which an allowance has not been made in the main pipe. It is true that the manufacturers of blowers advocate the use of cut offs, or did when they put out catalogs descriptive of blow-pipe work, but it has been amply proven that the indiscriminate use of them is a detriment rather than a benefit.

To illustrate, let us suppose that the machine in

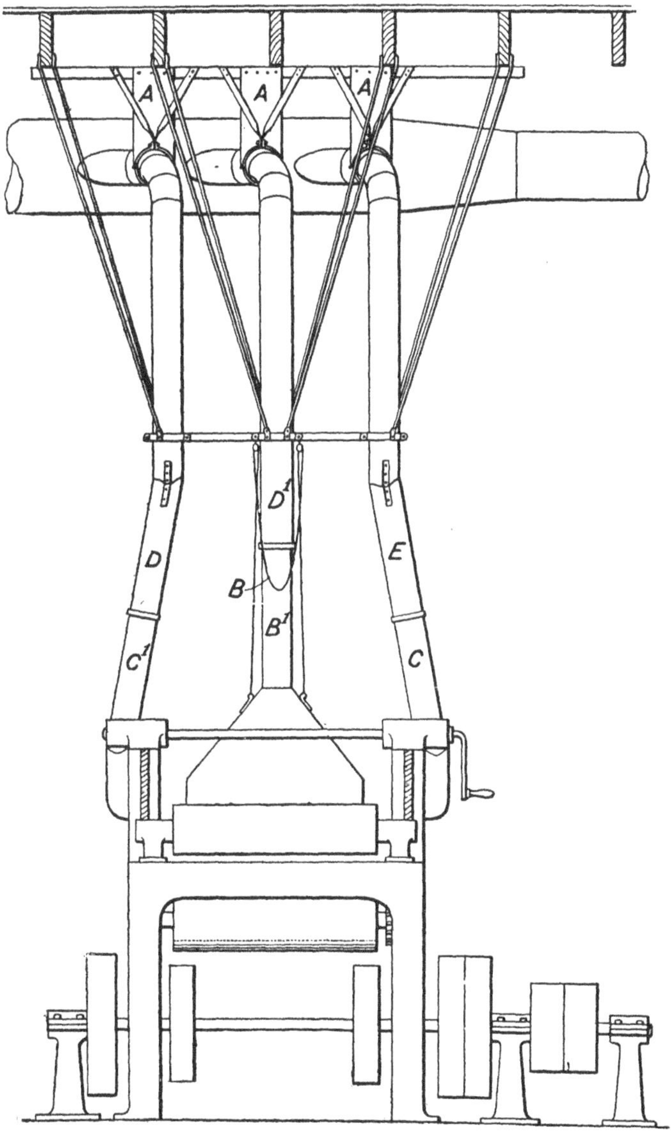

FIG. 13—PIPE CONNECTIONS FOR A FLOORING MACHINE.

Fig. 13 and another of the same capacity side by side are both shut down and all gates are closed. That means that 165 square inches of area, practically equalling the capacity of an 14½-inch pipe, has been shut off. On the average job that would mean considerable more than one-half the capacity of the main pipe at the fan.

The suction, while greatly increased at the inlets remaining open, would be weakened in the big main to such an extent that the material will invariably collect on the bottom, not only in the main suction, but in any elbows that may be in the discharge pipe, and no amount of pounding on the outside of the pipe or speeding of the fan will move them once they become packed in the curve of an elbow.

It may be argued that judgment should be used in closing these gates, that enough should be left open at any point in the plant to insure the entrance of air enough to clear the big pipe. This is good argument perhaps, but the average machine operator is lacking in judgment in matters involving such intricacy, which is the chief reason for his being a machine operator.

The best argument in favor of "cut off" is that every pipe cut off reduces the power required to drive the fan.

This is a fact, but where is the saving?

The coal bill argument falls flat here, because the average mill doesn't burn it. Their problem is not to save fuel, but to consume it. If the mill has not power enough to drive the fan wide open the job will be a failure whether cut offs are used or not.

In Fig. 13, A, A, A show the clamp braces attached to the horizontal pipes. Fig. 14 shows an enlarged view of one of these braces.

The clamp A in Fig. 14 is made of 1 by ⅛-inch band iron, as are also the two straps B and B¹. The straps are held to the clamp by a ¼-inch bolt, and are "draw nailed" to the 2-inch piece D, which is nailed to the joists above, as shown.

The board C is fitted to the pipe as shown, and also nailed to the 2-inch piece D. The straps are then drawn tight by pointing the nails slightly upward.

The manner of guying these uprights is shown in Fig. 13. The center or top head pipe is first securely guyed four ways; the two side pipes are then strapped to it as shown, and guyed two ways on the outside. For this purpose 5/16-inch bar iron should be used, the straps being of 1 by ⅛-inch bar. This not only makes a neat appearing job, but is perfectly rigid, as it should be.

A lifting cord for the top hood is shown at B in Fig. 13. The cord is run through two awning pulleys attached to the lug band, and fastened to the hood on each side. This is to facilitate the raising of this hood, which at best is an awkward job, since the hood is difficult to reach.

B¹, C and C¹ are slip joints, which are necessary for the raising and removing of the hoods, also for the telescoping of the side heads, as described in a previous chapter.

In Fig. 16 are shown the swinging or knuckle joints. The hinges are made of two pieces of 1 by ⅛-inch bar, the upper one being offset and fitted over the lower one to form a flat surface against

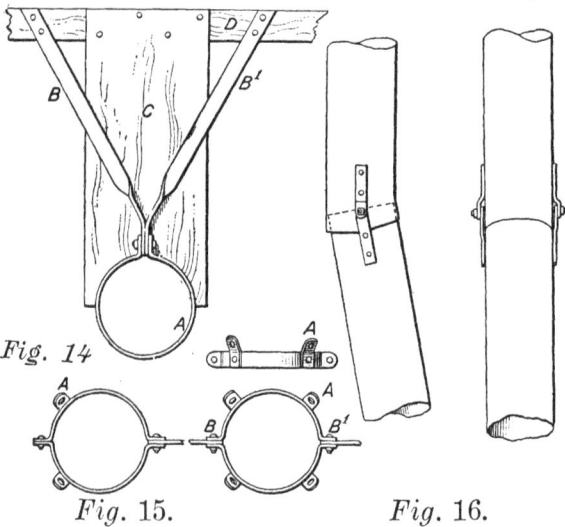

Fig. 14

Fig. 15.   Fig. 16.

DETAILS OF CLAMPS AND JOINTS.

the pipe. The lower one is countersunk on the under side for a screw head stove bolt. The upper joint of the swing is expanded on opposite sides, as shown, in order to allow the lower joint to swing. Care must be taken not to expand the joint too much.

The "lug bands," shown in Fig. 15, are also made of 1 by ⅛-inch bar. Bars of this size are used to clamp the galvanized sheets together, and

are consequently plentiful in the shop and ideal
for the purpose. Certainly, if one is regularly engaged in the blow-pipe business on an extensive
scale it pays to have the lug bands cast; they must,
however, be malleable. The bands shown in Fig.
15 are made with a draw on each side, as shown
at B and B¹, and a ¼-inch bolt is used to draw
them together. The guy lugs, shown at A, are
flush, riveted to the side of the band.

D D¹ and E in Fig. 13 are sleeves, which consti-
tute the outer section of the telescope. These are
stiffened at the lower end by a 1 by ⅛-inch band,
flush riveted to insure easy working of the slip.

CHAPTER VII.

DESIGNS FOR HOODS AND SWEEPERS.

The present chapter will deal with the design and application of saw hoods and floor sweepers, also a shaper hood.

In Fig. 17 is shown a hood attached to a band saw, together with an enlarged view of the same. In the enlarged view the door is shown open. This door is necessary, as without it the saw could not be taken from the wheels. A piece of wood W, is fastened along the bottom of the hood inside, having a slot through which the saw runs. This hood is very efficient since it is not enlarged to any extent above the diameter of the outlet pipe—or suction. The suction at the saw is therefore very strong.

In Figs. 18 and 19 are shown two styles of circular rip saw hoods. These hoods may, of course, be used for crosscut saws as well, both being table saws. In the writer's opinion No. 1 is the better. It is cheaper than the other and practically answers the purpose, but, theoretically speaking, No. 2 has the best of it.

These hoods must be made wide enough to permit of the saw being taken off the mandrel and have the side cut out as shown, so that the nut and washer may be easily removed. The pipe connections must, like all bottom hood connections, be made removable to facilitate cleaning out.

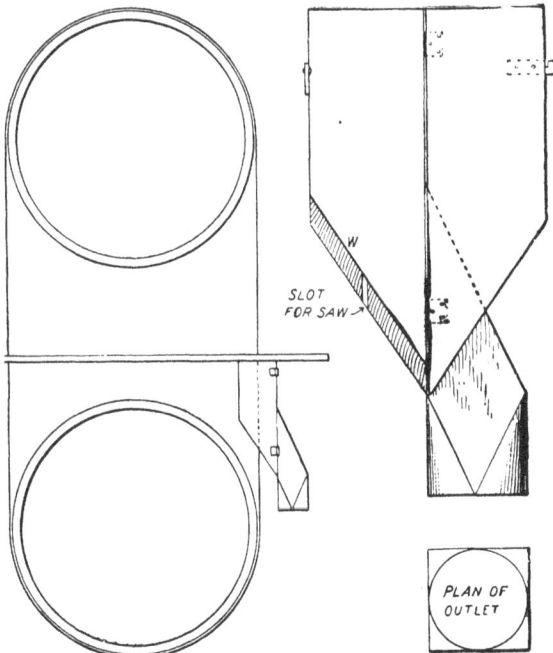

FIG. 17.—BAND SAW HOOD.

In Fig. 20 is shown the manner of handling a circular re-saw. It will be seen that a large door is placed in the side of this hood, which must be made large enough to admit of the saw being slipped off the mandrol. It is a two-man job to remove this

saw and the hood must be entirely out of the way.
This hood should be made of extra heavy material
since the centrifugal force of the saw is very great

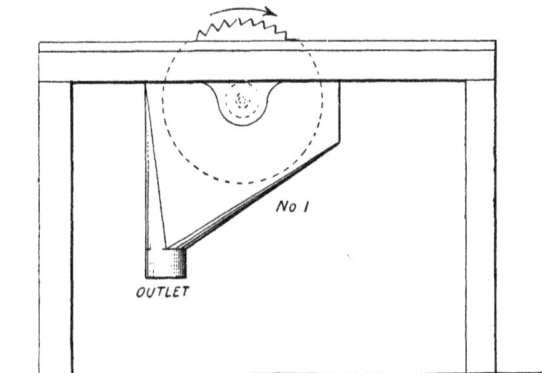

FIG. 18.—RIP SAW HOOD.

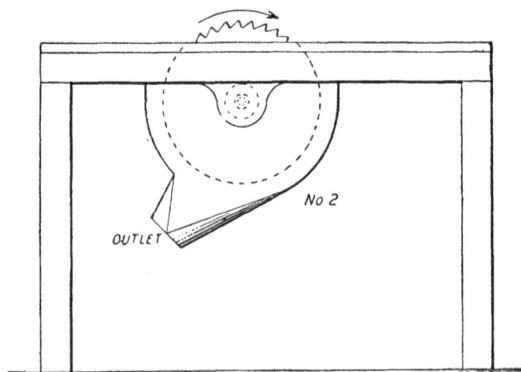

FIG. 19.—ANOTHER FORM OF RIP SAW HOOD.

and the material thrown off is coarse and heavy.
In Figs. 21 and 22 are shown designs for a top
and bottom floor sweeper. By top floor sweeper

is meant the sweeper that must be placed on the upper floors of a mill if the main pipe is underneath.

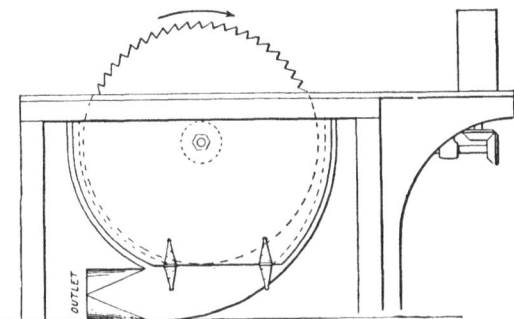

FIG. 20.—CIRCULAR RE-SAW HOOD.

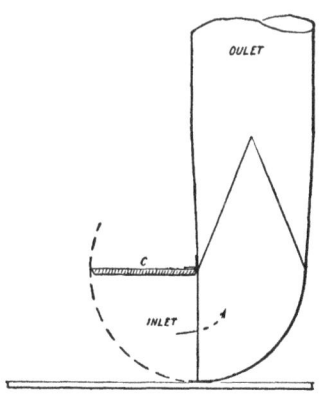

FIG. 21.— MAIN FLOOR SWEEPER.

It will readily be seen that if a simple floor flange is used to sweep into, nails, nuts, bolts, or any heavy foreign matter would be swept into the pipe. At A in Fig. 22 is shown an angle plate riveted inside the hood. The material is never swept into the hood, or should not be, but is pushed near enough for the suction to take it up. The angle plate necessitates the lifting of the material, hence anything heavy enough to do damage will not get over it.

Fig. 21 is the common type of floor sweeper, the inlet being rectangular and not much in excess of the size of the pipe. The doors shown at B and C in both sketches must be closed when not in use, as it is not the rule to allow for floor sweepers in figuring capacities.

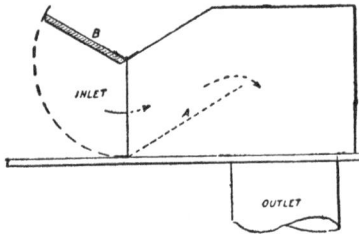

FIG. 22.—TOP FLOOR SWEEPER.

I do not think it is good practice to flare out the inlets of floor sweepers, because the broom artist is thus encouraged to pile the shavings up in front of it and see how much he can make it swallow. He generally chokes it and has to clean it out for his pains.

Fig. 23 shows the design of a hood for a shaper, or, as it is sometimes called, a variety molder. This machine is simply a cutter head, on the end of a perpendicular shaft, sticking through a table.

The operator may work at it from any position. For this reason the hood is made to revolve at A, the pipe being exactly over the center of the shaft.

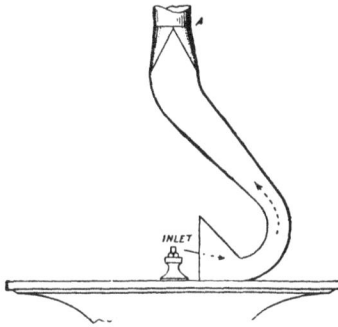

FIG. 23.—SHAPER HOOD.

The hood is thus always the same distance from the cutter and may easily be placed in the position wanted.

CHAPTER VIII.

Hoods for Special Machines.

I have described in previous chapters the hoods most commonly used, with a view of acquainting the student with the basic principle of hood construction in general. It would be a difficult matter to set before the reader in detail the many different styles of hoods necessary, for instance, in a wagon or carriage factory, where axle lathes, drum sanders, belt sanders, disc sanders, double cut-off saws, tenoners and many more or less complicated machines abound, and where the material handled is selected from hard wood dry and light. I think it is hardly necessary to do so and am a little afraid of befogging the reader's mind with too much description.

I propose to deal in this chapter with the application of the principle already described as adapting all hoods to varying conditions. This line of endeavor calls for a high degree of intelligence, and I will say in passing that in no branch of it is intelligence and ingenuity a greater factor than in the hooding of machines.

I have described in a previous chapter how a flooring machine should be piped. The same procedure is necessary with molders and stickers. A flooring machine, when running steadily, will knock off a good sized wagon load of shavings in an hour, yet such a machine, when properly piped and hooded, will not drop a good sized keg full of shavings in two hours of cutting.

Molders and stickers, owing to a difference in construction, as well as to the variety of work they do, are more difficult to pipe successfully, and indeed the results obtained with the flooring machine will not be expected of them, and certainly should not be promised.

The planing mill and sash and door factory are really the easiest plants to pipe successfully and are good places for the beginner to get his initial experience. Wagon factories, furniture factories and plow factories may be placed in the same category.

The stave mill or barrel factory is in a class by itself and is perhaps the most difficult of all woodworking concerns to pipe. This being the case, let us tackle it now. There are in all stave mills what are known as stave jointers and also heading jointers. In these machines the knives are on the side of a large wheel. The operator stands before it, clamps the stave or heading "bolt" upon a carriage, pushes it against the rapidly revolving wheel, when a shaving is taken off the full length of the stave, if the "bolt" is a little green, which is often the case. Thus it will be seen that this plant requires what is known as a long shavings fan.

The long shavings fan has no rim on the wheel and consequently nothing for the shavings to catch on.

The stave, or heading jointer, however, is a fan in itself, the whole being encased in sheet iron, with an outlet the same as a fan. The inner portion of the knife acts as a fan blade and the center of the wheel is the air inlet. Such wheels will take care of their own shavings for a distance of not more than 150 feet on a horizontal line, but there must be no elbows in the pipe, except the one leaving the outlet, if the outlet is perpendicular, and there is no perpendicular rise in the pipe.

If the boilers or shavings bin are too far away from these machines, we must run them independently into a separator and connect the separator into the main suction of our fan, always near the fan itself. Never blow air into a main suction.

There are also in this mill what are known as equalizing saws, two saws run on the same shaft, which saw the stave or the heading bolt the proper length, sawing both ends at the same time.

The hood described in Chapter VII, attached to a circular re-saw, is the hood to use here, but it should be made of No. 12 steel. The door should be flanged in along the top between the hub of the saw and the cutting edge, as close to the saw as possible, to prevent thin blocks from entering the pipe.

If tight barrels are made in this shop we may be asked to pipe what are known as barrel lathes. Don't do it. The barrel is set in a machine which clamps it at each end and causes it to revolve rapidly. The operator applies a tool to it some thing on the order of a spoke shave. He produces an abundance of splinters, which are perfectly straight and about the length of a toothpick. These can only be successfully handled with a wheelbarrow or some sort of belt conveyor.

To depart for a moment from the subject of hoods, while discussing a stave mill proposition, I wish to impress the reader with the fact that a job in such a plant must be constructed of especially heavy material and must be solidly erected.

There must be no crotches, projecting edges, or even high rivets in any pipe for long shavings to cling to, and the material outlets of all separators must be made especially large.

Spiral pieces, such as described by M. F. H., in Chapter XIV, and by the writer in a chapter foregoing, must be put in the cones to stop "whirling," or the long shavings will invariably clog the machine. So much for a stave mill or cooperage works.

In wagon shops and furniture factories the work is pretty much alike. The wood, as I have said, is dry and hard and there is a great deal of sander dust to handle. Now a belt sander is not an easy thing to hood successfully.

It is nothing more nor less than a sandpapered

belt running over two pulleys, and is used for sanding spokes, felloes, chair rounds, legs, etc. Do not make the mistake of using a wide flaring hood for this, but so construct the hood that the suction will be thrown near the belt at the point where the dust leaves it.

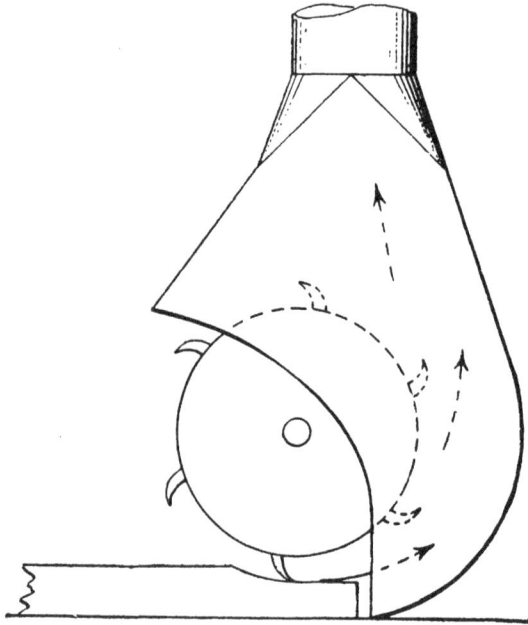

FIG. 24.—THE RIGHT WAY.

A wide flaring hood is always a waste of material, unless the shavings are thrown directly at it.

In this factory will be found a double cut-off saw. These saws are like the equalizers in a stave mill, with the difference that the saws can be moved apart or brought together to saw pieces of varying lengths. .

The hood for the movable saw, therefore, must be connected with a telescope joint if the pipe is beneath the floor, and with a swinging telescope if the pipe is above it.

In either a furniture or wagon shop will also be found several tenoning machines. These ma-

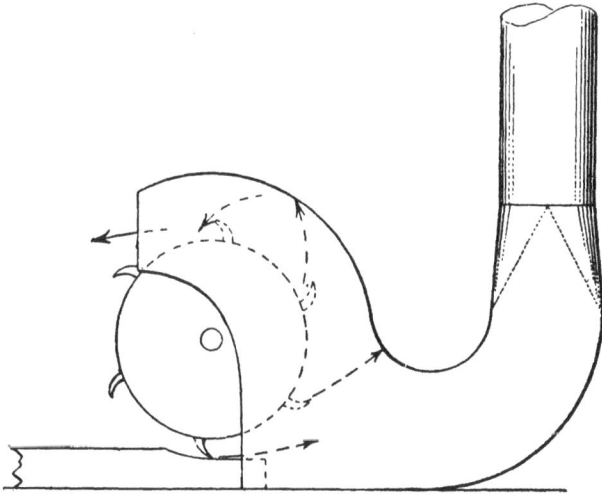

FIG. 25.—THE WRONG WAY.

chines knock off a light curly shaving which is generally very large. For this reason a large pipe should be used, more for the sake of room than anything else, as a strong suction is not needed. The proper hood for this machine will readily suggest itself, as there is only one way of hooding it.

It is well to remember always, that the outlet of any hood should be so placed, if possible, that no chips will fly past it to be deflected back on the knives and thereby thrown out of the mouth of the hood.

The illustrations show the wrong way and the right, and generally the most practical way.

It is not always easy to tell by watching the rotation of a cutter "in the wood" the direction taken by the bulk of the shavings. I have found it good practice to hang a board in front of the knife and note where the chips hit the board.

There is no work more interesting than that of studying out hood designs and none requiring more originality.

Don't be afraid to experiment, and don't depreciate any advice from the "presiding genius" of the machine.

CHAPTER IX.

We have spoken of blowing the material from stave jointers into a separator independently, and thence into a main suction near the fan.

If one or more pipes are connected to a separator, care must be taken that they are not so constructed that the material will pack in the inlet.

Fig. 26 shows a plan and elevation of a cyclone separator in which, at AAA, three small pipes are shown entering the inlet. If for any reason the bottom one of these three pipes is not working, if the jointer is shut down for instance, the light dust from the top pipes will gradually back into it and, in a very short time, it will become clogged. For this purpose a special separator should be made with each pipe tapped into it independently, as shown in Fig. 27. This is nothing more or less than three pipes joining a cylinder tangentially, each pipe being at right angles to the axial line E.

The reader may think that this is rather a simple looking contrivance, perhaps too simple

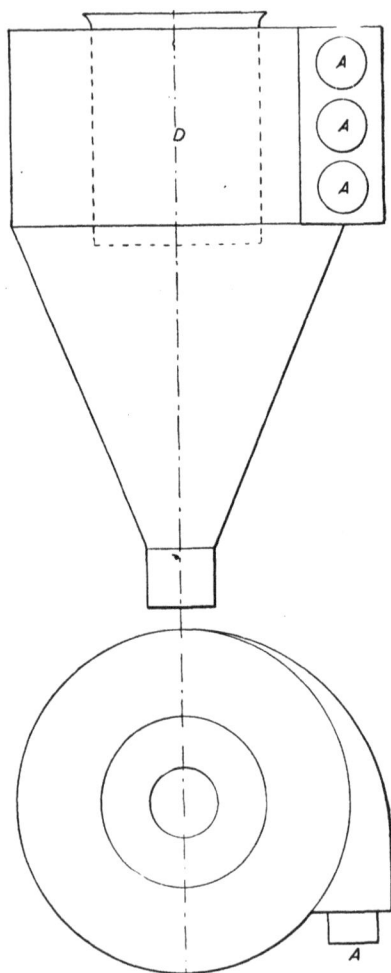

FIG. 26.—SEPARATOR WITH PIPES ENTERING AN INLET.

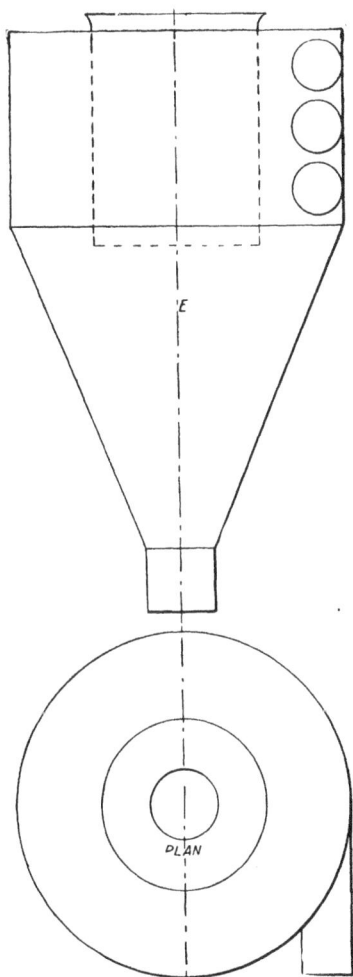

FIG. 27.—SEPARATOR WITH PIPES ENTERING TANGEN-
TIALLY.

to be effective. Let me say in passing that he who has traveled much and has had an eye out for such things, has seen many different styles and shapes of separators. While it is not my business to criticise any make or design of separator, I think it but just to assure those who have wondered why some are so oddly shaped and complicated, that the person with the simple pattern got on the market first and the other fellow had to complicate his to get it patented.

The principle involved in a "dust arrester" which, by the way, is the proper name for it, is centrifugal force and nothing is needed to create this but a cylinder with a tangential inlet, a cone at the bottom of the cylinder for the purpose of collecting the material and depositing it after the wind has left it, and an outlet for the wind which should be at least three times the area of the inlet, with a tube reaching down an inch or so below the bottom of the inlet, as shown at D in Fig. 26.

Now, of course, the matter of proportions of the body of this machine should be considered. The inlet of it, except in such special cases as those just mentioned, should be rectangular and something in excess of the round pipe that enters it, the excess being enough to overcome the friction due to transforming. Suppose, then, we are making one for a 24-inch pipe, the inlet (C in Fig. 28) should be 16 by 30 inches; the wind outlet A, 42 inches, about three times the area of 24 inches; the shavings outlet B, 14 inches; the

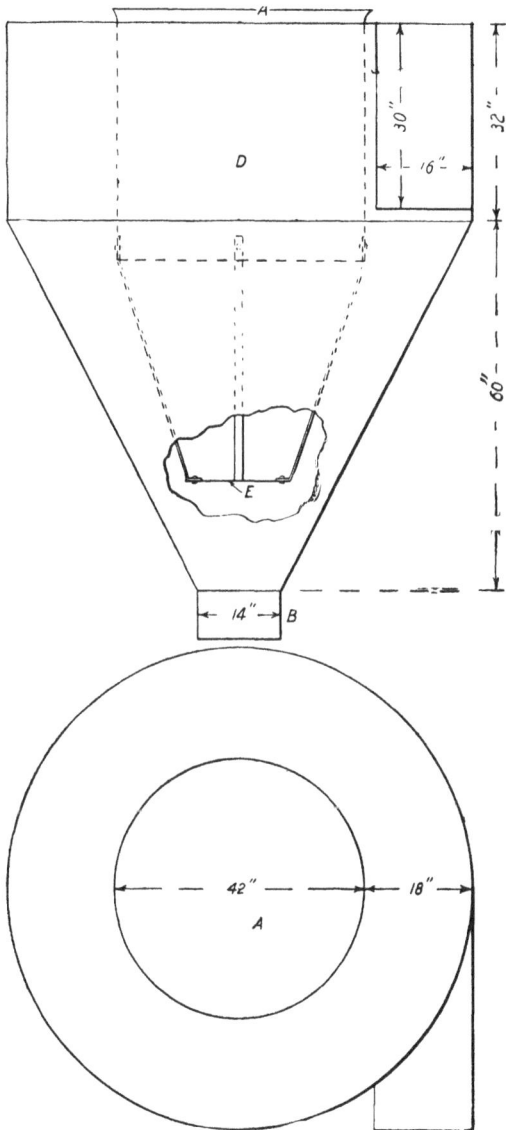

FIG. 28.—CORRECT METHOD OF CONSTRUCTING SEPA-
RATOR.

shell diameter, 6 feet 6 inches: hight, 32 inches; hight of cone, 60 inches.

At E in Fig. 28 is shown a disc fastened to the tube D by braces, shown by dotted lines. The disc should be made somewhat larger than the shavings outlet. Its purpose is to prevent the material from being caught in a center current and carried upward.

This current is apt to form if the disc is not there. It should be placed so as to leave a space of from six to eight inches between its edge and the side of the cone. If a spiral is put in the separator the disc will not be needed.

I will assure the reader at this point, that a dust arrester, made in the proportion here given, will work as close, by which I mean, lose as little dust, as any machine of its kind ever built. It is no better and no worse than the other successful types. It may be made larger and still do as well. It should not be made much smaller.

To touch again upon the subject of depositing material from a separator into a fan, I will say that this practice gives excellent results in a large mill, where three or four fans are used, and the distance to the fire-room is so great that discharge pipes from each fan to the boilers would be an expensive proposition. Collect the material from the fan farthest away into a separator and deposit it into the next fan; the same with the next, and so on, letting the fan nearest the furnaces carry the whole product to the separator at that point.

Now, with a good suction pulling at the bottom of the separator, the pipe need not be larger than from 6 inches to 9 inches, according to the size of the machine and the amount of material handled. This should never be practised, however, in a long shavings plant, and it is rather risky in a planing mill, because there is too much material to handle. But it is an old scheme in sash, wagon, and furniture factories and the like.

CHAPTER X.

EFFICIENCY OF THE EXHAUST FAN.

I am indebted to one of the well-known blower concerns for the capacity table printed in this chapter. This is presented for the reason that the power required to drive exhausters is an important factor, when a deal is being negotiated in the piping business. Yet it is a factor very often regarded as one of small importance.

By referring to this table the reader will see

No. of Fan	Diameter of Wheel, Inches	Width Periphery Inches	Diameter Inlet [Inside] Inches	R. P. M.	½-Oz. Pressure Cubic Feet per Minute	Brake Horse power
25	16	6⅛	10	985	1095	0.30
30	19	7⅛	12	830	1580	0.43
35	22	8⅛	14	715	2155	0.59
40	25	9⅜	16	630	2820	0.77
45	28	10⅞	18	563	3560	0.97
50	31	12⅜	20	508	4400	1.20
55	34	13½	22	464	5330	1.45
60	38	14½	24	415	6350	1.73
70	44	15⅛	27	375	7440	2.02
80	50	16½	29	328	10050	2.75

No. of Fan	¾-Oz. Pressure			1-Oz. Pressure		
	R. P. M.	Cubic Feet per Minute	Brake Horse power	R. P. M.	Cubic Feet per Minute	Brake Horse power
25	1200	1345	0.56	1390	1555	0.85
30	1012	1940	0.80	1170	2240	1.22
35	876	2635	1.08	1010	3040	1.66
40	772	3450	1.41	890	3980	2.17
45	689	4360	1.78	795	5030	2.74
50	622	5390	2.20	719	6220	3.39
55	567	6525	2.66	655	7530	4.10
60	509	7775	3.18	587	8960	4.89
70	459	9120	3.72	530	10500	5.72
80	402	12100	4.94	464	13980	7.62

No. of Fan	1½-Oz. Pressure			2-Oz. Pressure		
	R. P. M.	Cubic Feet per Minute	Brake Horse power	R. P. M.	Cubic Feet per Minute	Brake Horse power
25	1710	1910	1.56	1966	2200	2.40
30	1436	2750	2.25	1655	3175	3.46
35	1240	3730	3.06	1430	4310	4.70
40	1090	4880	4.00	1260	5640	6.15
45	976	6180	5.06	1125	7140	7.79
50	882	7640	6.25	1015	8820	9.63
55	805	9250	7.58	927	10650	11.60
60	720	11000	9.20	830	12700	13.85
70	650	12900	10.57	750	14875	16.20
80	569	17170	14.05	656	19800	21.60

No. of Fan	3-Oz. Pressure			4-Oz. Pressure		
	R. P. M.	Cubic Feet per Minute	Brake Horse power	R. P. M.	Cubic Feet per Minute	Brake Horse power
25	2420	2690	4.40	2780	3100	6.78
30	2035	3875	6.32	2345	4475	9.77
35	1755	5260	8.60	2025	6085	13.26
40	1545	6885	11.22	1785	7955	17.35
45	1380	8710	14.36	1590	10050	21.90
50	1247	10740	17.50	1440	12420	27.10
55	1138	13000	21.20	1310	15050	32.80
60	1020	15500	25.25	1175	17900	39.04
70	920	18150	29.60	1060	21000	45.80
80	805	24200	39.50	930	28000	61.00

Double fans have double the capacity given above and require twice the power to drive at given speeds and pressures.

This table is based upon the inlet and discharge pipes being the same area as the fan inlet. If the suction area is less than the inlet of the fan, the volume and horse power will be reduced and the pressure increased.

TABLE OF

SPEED, CAPACITY AND HORSE POWER REQUIRED FOR

STEEL PLATE EXHAUST FANS.

how increasing the speed of a fan by a few revolutions will more than double the amount of power required to drive it. Take, for example, the 40-inch exhauster fourth in the lower table: 4 horse power will drive it 1,090 revolutions per minute, yet to drive it 1,785 revolutions, an increase of speed of but 695 revolutions, requires 17.35 horse power. The reader will note also the last statement made underneath the table, viz.: "If the suction area is less than the inlet of the fan, the power and volume will be reduced and the pres-

sure increased.'' Thus, if it is a question of
power with the prospective purchaser, sell him a
larger fan.

I am indebted to the same concern for another
table, given below, and which shows how speed
can be cut down and power saved by adopting the
suggestion.

To quote the American Blower Company: Suppose we have 284 square inches of area in all the
branch pipes and the main suction pipe after the
last branch is taken in—19 inches in diameter.
The various sizes of fans which can be applied,
with their respective results, are shown in the
table below, this being based on 100 feet of suction pipe, 100 feet of discharge pipe, four elbows
in the pipe and a properly proportioned separator:

Size of Fan	Speed	Horse power
45 inches	1,300 R. P. M.	11⅔
50 "	1,010 "	8¾
55 "	810 "	7
60 "	650 "	5⅔

Thus it will be seen that to use a 60-inch fan instead of a 45-inch is to reduce the power more
than one-half. I beg to remind the reader that
these deductions are the result of scientific tests
and are indisputable.

Another factor to be considered here is wear
and tear. It is obvious that a 45-inch fan running
at the rate of 1,300 revolutions per minute will

wear faster than a 60-inch running at the rate of only 650 revolutions per minute. It is not only the difference in speed and consequent reduction of friction that counts, but the larger fan is heavier.

There is another fact connected with these fans and, indeed, with all exhaust fans that have a standing, which it is well to know. The bearings are self-oiling and the oil chamber is very large. The man who is installing a piping system should see to it that these chambers are filled with good lubricating oil before the fan is started.

In previous chapter I stated in effect that there may be a question of patent rights on dust arresters. It may not be generally known that there are at least two makes of dust arresters still protected by patents. The old style "cyclone" pattern, as it is called, a sketch of which appeared in Chapter IX, may now be built by anyone.

The cheapest way to build these machines is to construct the top of wood, sawing it out in segments. The wood should be seasoned pine about 1 inch thick, or ⅞-inch dressed. Cover both sides with No. 24 galvanized iron, being sure to have no laps near the joints in the wood. It is not necessary to nail the wood segments together as the lining will hold it rigidly. Then nail this top to the shell, which should be made of heavy galvanized iron, since it is subjected to a great deal of friction. The cone need not be so heavy but may be made one gauge lighter. This should be riveted to the shell and the whole should be either soldered or cemented inside.

The top, if made of wood, should be well soldered to protect it from the weather, which would otherwise soon rot the wood. With regard to cementing these separators inside, there may be some who will question the advisability of this procedure.

I will state here that if one has a fan with a large amount of discharge pipe, it will pay to build the whole of black iron, separator and all. Common putty, if forced in behind the laps inside, will remain there and perform its function as long as the job lasts.

Recently the writer saw such a job which he had installed nine years ago, and which to all appearances was good for nine years more, putty and all.

CHAPTER XI.

USE OF THE TWO-WAY MIXING VALVE AND THE AUTOMATIC DAMPER.

In lumber plants, by which I mean the mills where lumber is made from the standing timber ready for the market, it is often the case that the sawmill is situated at a distance of from five hundred to a thousand feet from the planer on which it is to be dressed.

I have never blown shavings a thousand feet, and I really doubt if it can be done successfully. In a straight pipe that long the material will arrive at the suction end of the job if the blower is kept up to speed, but I doubt if the ordinary results will prevail.

I know, however, that they may be blown a distance of six hundred feet in a comparatively straight line with fairly good results all around. Of course, I am assuming that a separator is used at the end of the pipe in both cases.

Suppose we were piping a large planing mill and using a double 60-inch fan, and suppose the purchaser wanted part of the product delivered at his sawmill six hundred feet away. Ordinarily a gate valve would be put in the discharge pipe somewhere near the fan having a 35-inch inlet and two 35-inch outlets, one for the sawmill and one for the planer. Of course this would be a poor expedient, for it is obvious that while the sawmill is being supplied no shavings could be fed into the furnace of the planer. The gate could not be set half-way because the fan only produces wind enough to clear the pipe, and if half of this were thrown into another pipe neither of the pipes would "clear." If an ordinary Y-branch were used, of course the shavings could not be controlled, and there would be times when practically no shavings would be wanted at the sawmill.

The mixing valve is a valuable device for two reasons—it will throw 85 per cent. of the shavings either way and by its use a pipe just half the size of the fan outlet is all that is needed to deliver the goods either to the planer furnace or that of the sawmill six hundred feet away.

When we consider the difference between the cost of six hundred feet of 35-inch pipe and a separator large enough to handle it and the same amount of 25-inch pipe and a smaller separator, we find that this is no small item. Again, we might have one fan in the mill and two separators both at the mill, one over the furnace and one over a storage vault for shavings that are sold and hauled away in wagons. Here the mixing valve is necessary again, for with its use two small separators can be built, the combined capacity of which would equal that of the fan. Without its use each separator would need to be of a capacity equal to the fan.

In Fig. 29 is shown a plan and side elevation of the mixing valve. The inlet A has the capacity of the fan. The outlets E and E have each half the capacity of the fan. It will be seen that the inlet A is a "square to a round" which is connected to a transforming piece made rectangular at G. The area of this rectangle should be a little in excess of that of the square at the inlet to eliminate friction as much as possible and yet maintain the pressure.

The width of the rectangle should not be more than half the width of the square. The valve when in operation must set as it is shown in the side elevation. From the rectangle G the valve is again transformed to a square, as shown, equal in area to the inlet A.

The top of the first transformer, shown at B, serves to baffle the air, throwing the current downward. The shavings, of course, will follow, and the bulk of them will remain traveling along the bottom until they strike the lower gate at D.

It will be seen that the valve has two gates, C and D; the dotted lines on the lower gate in plan show a guide, H H. This is a flat piece of heavy iron fastened to the top of the gate to serve the purpose of deflecting the shavings that may bound upward through impact on the gate and

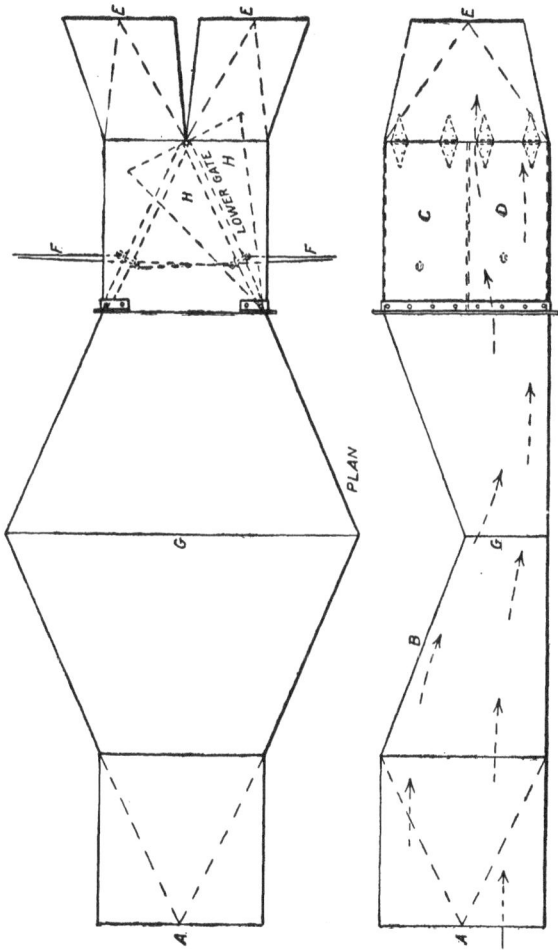

E

E

F   H   F

LOWER GATE   H

G

PLAN

A

E

C

D

G

B

A

SIDE ELEVATION

FIG. 29.—MIXING VALVE.

throw them into the outlet. If the lower gate is thrown to the right the upper one must be thrown to the left and vice versa, in order to equalize the air.

Thus it will be seen that while the bulk of the material may be thrown either way the air is never disturbed, but will flow equally through both outlets.

It is needless to say that this valve must be solidly constructed. The gates are operated by wire ropes F and F, secured to the gates by collar

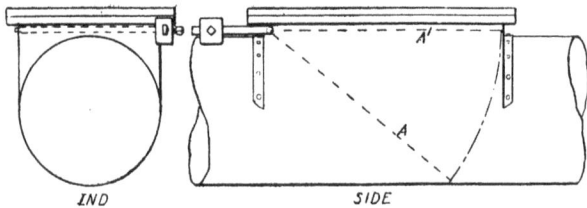

FIG. 30.—AUTOMATIC DAMPER.

eyebolts, as shown. When the gates are pulled past the center, the wind will slam them against the sides of the valve with considerable force. For this reason the gates should be made of wood sawed pointed at the ends, as shown in the plan, and made thick enough to stand the strain and lined with heavy iron. They can be hinged to the crotch sheet, as shown in the elevation. The sides where the gates strike should be reinforced with angle iron, bent over the top and bottoms, as shown, and securely riveted.

THE AUTOMATIC DAMPER.

The automatic damper shown in Fig. 30 is a fire preventative and should be installed as near the separator as possible. The illustration shows a side and an end view of the device which is nothing more nor less than a joint of pipe the size of the discharge with a box riveted to the top of it, long enough to receive the damper A, A¹ when the wind is holding it up.

The counterbalance B should be so placed on the handle that the damper will barely drop of its own weight when the fan stops. This damper prevents any fire from the furnaces getting back into the plant. A wooden frame is shown around the top of the box to stiffen the same and to hold a cover which may be screwed on but made easily removable.

CHAPTER XII.

In Fig. 31 is shown a piping system as applied
to a forge shop.

The piping, indicated by dotted lines, is under
the floor of the shop and should be made of tile.
This is the suction which carries away the smoke
and gases emitted by the forges.

In view of the fact that in large shops of this
kind swinging and traveling cranes are in con-
stant use, under the floor is the only place to put
the pipe, since it would be in the way overhead.

The fan, which is a low-speed exhauster, in-
tended more for volume than pressure, may be
set on or near the floor in an out of the way place
and connected to the tile with sheet iron elbows.

The discharge should run out through the roof
in a straight line, if possible, and a weather cap
be put on the top of it. Care should be taken in
the construction of this cap to make it weather
proof.

An elevation of the canopy or hood, which is
designed to catch the gases, is shown in Fig. 32.

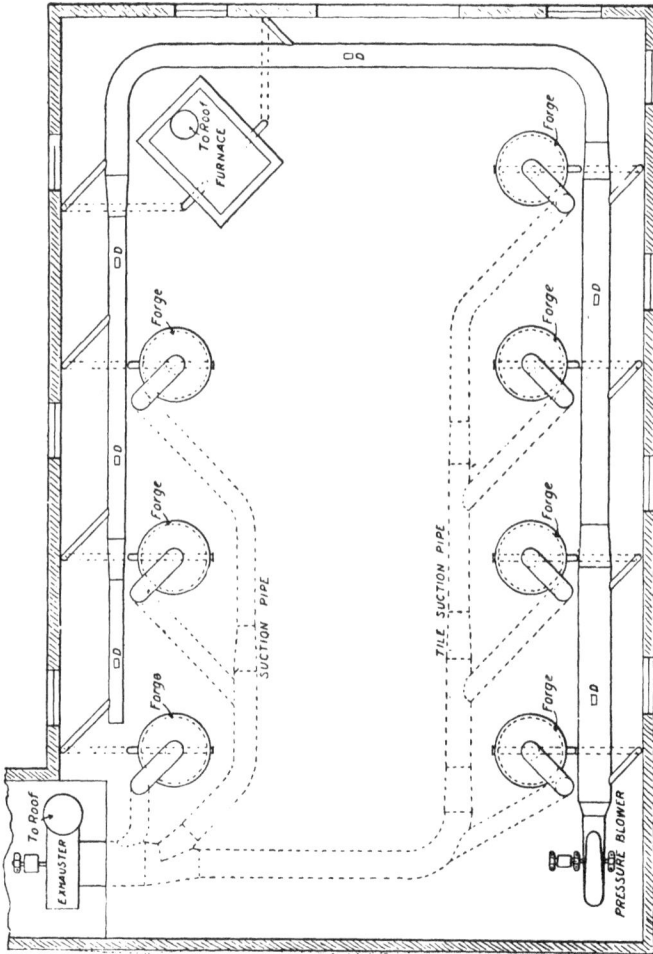

FIG. 31.—PLAN OF FORGE SHOP, SHOWING BLAST AND SUCTION SYSTEMS.

The hood is practically a double one, as shown, the outside hood having a 10-inch outlet, while the inside one need not have an outlet larger than 4 inches, the object being to throw the draft to the outside of the hood, as indicated by the arrows. For this purpose the inside hood is made 3 inches smaller all around than the outside and connected to the same by the use of bolts and ferrules, as shown in the small detail at the right.

If a single hood were used in this case, the draft, while strong at the apex of the hood, would be scarcely perceptible at the bottom and very little of the smoke would be influenced by it. By using the inside hood the draft, as before stated, is thrown to the outside to a large extent, and the hood becomes more effective.

The forges shown in Fig. 31 are "drum" forges, being cylindrical and made of about 3/16-inch steel, and having a 4-inch wrought pipe run through them with a plug at one end for cleaning out, the blast pipe being connected at the opposite end.

At the center is the tuyere, a cast iron box perforated at the top, the wrought pipe being connected at each side of it.

The size of this box varies according to the size of the forge. If a flange fire is needed the box is made long. A large square box is used for a welding fire, and a small box for ordinary forging.

A 4-inch blast pipe is large enough for any forge where ordinary work is done. As the smith stands at the forge his anvil must be at his right, his blast-gate lever at his left. The blast-gate lever is shown at B in Fig. 32.

In a corner of the shop, Fig. 31, is shown a fur-

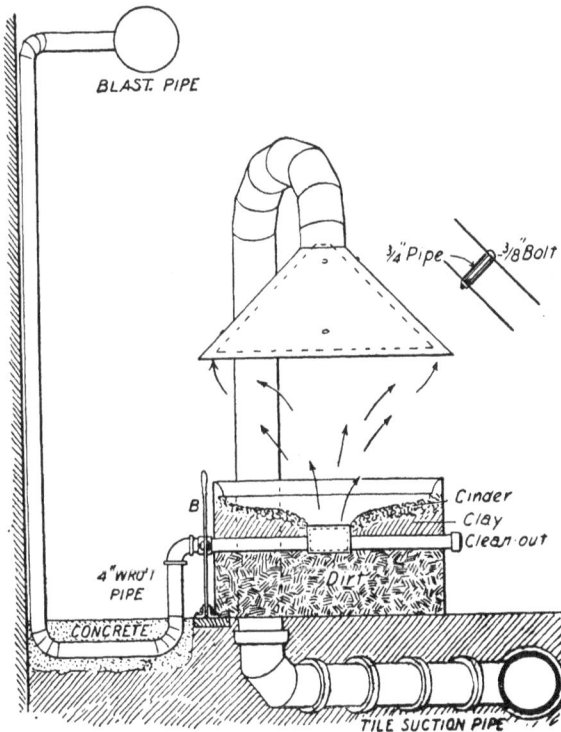

FIG. 32.—SECTIONAL ELEVATION OF FORGE.

nace which is used for heating large work for forging under the steam hammer. This furnace should have a 3 or 4-inch blast pipe on each side, according to the nature of the work.

It is needless to say that the blast pipe cannot run across the floor of the shop, which will either be cement or cinders, generally the latter. Consequently, if the forge or furnace sets away from the wall, they must run under the floor. It is a good idea to dig a trench for these pipes, box them on the sides and cover the whole with concrete. Of course the trench should be considerably larger than the pipe—say 3 inches larger all around.

At the end of the pipe next to the forge, the connection should be wrought iron, since if the sheet iron pipe were brought up it would soon be battered out of shape by tools and iron dropping from the forge. The pipe which is run beneath the ground need only be slapped together if it is to be concreted, since the concrete when set will hold it rigid.

It will be seen in Fig. 31 that the main blast pipe follows the side of the building clear around the shop. This is done for several reasons; one reason is to keep out of the way of cranes, another is the forges are generally near the wall, and the branch pipes to them may therefore be run down the wall and fastened to it. Thus they are out of the way besides being fastened securely.

The method of erecting this pipe is the same as the suction or the same as blow piping, excepting the fact that the laps are made the opposite way, the wind traveling toward the small end of the pipe instead of toward the large end.

In Fig. 33 is shown a damper, which is controlled by the blast, swinging down when the fan is not running and closing again when the blast is on. The use of this damper is to ventilate the pipe when the plant is shut down, since gases from the forge will back up into the system at night, especially if some of the gates are left open, which is often the case.

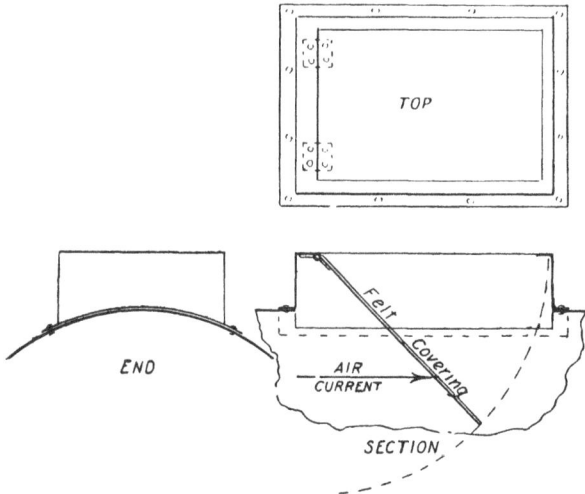

FIG. 33.—DAMPER FOR BLAST PIPE.

The writer recently saw a blast pipe 24 feet long burst open the entire length, the gas in the pipe being ignited when a helper started his fire in the morning and before the fan was running. This cannot happen where the automatic damper is used. The opening of this gate should be 4 by 6 inches, and one gate used for every two forges, as shown at D in Fig. 31. The blower used for this work should be a pressure blower and must be kept up to catalog speed.

To obtain the size of the main at the fan, figure out the number of branches and add their areas together. Order a fan whose outlet shall be equal to the area of this pipe, then start with a pipe 2 inches larger in diameter, and use the larger one as a basis when deducting for outlets. Thus the main pipe at every point will be larger than the outlets beyond it, the velocity will be less, and a great deal of friction will be eliminated.

CHAPTER XIII.

"Don'ts" and "Don't Forgets" for Blow-Pipe
Mechanics.

Don't clean hot coppers on your overalls; have a dip pot in your kit.

Don't use a fire pot whose bail is pivoted on the sides instead of the front and back, when you are working aloft, as there is always danger of the coppers falling out.

Don't wait until the boss tells you to file the coppers.

Don't forget to wash the acid off a soldered seam.

Don't forget that dry wood dust sometimes explodes when ignited.

Don't put charcoal made of bark in the fire when working in a mill.

Don't forget that keeping a straight even point on a copper makes soldering easier and insures a neater job.

Don't forget that galvanized iron will "cut" acid.

Don't forget that a lump of salammoniac in the kit box will rust the tools.

Don't knock the other fellow's work when you

are sent to repair it; you may be talking to his cousins.

Don't forget that a fan running at a top speed will collapse a main pipe if there are not sufficient openings in it to prevent a vacuum.

Don't depend on the holes to keep a line of pipe straight. It may turn out to be an elbow.

Don't connect pipe on the floor if you are riveting it on a rail or mandrel. Connect it on the rail or mandrel.

Don't use a hollow mandrel to rivet pipe of larger than 6 inches diameter. Use a rail long enough to reach at least the length of a joint and a half.

Don't forget that unnecessary movements are tiresome and expensive.

Don't stand on the right hand side of the rail when riveting pipe. Stand with the pipe on your right as you face the bench. Keep the hammer in the right hand, the set in the left, and pick up the rivet with the thumb and finger of the left hand.

Don't forget that an elbow cannot be made round after it is riveted together.

Don't forget that the helper who is holding on rivets on the inside has troubles of his own. Have some set signal to let him know when you are done with the rivet.

Don't forget that slow, regular blows make it easier for the helper to stay on the rivet and expedite the work.

Don't forget that perpendicular steam escape

pipes should be erected so the steam will go against the laps; thus the condensation will run back down the pipe and not lie between the laps. This is a good way to erect a large smokestack.

Don't "swell" too much when you find out that you know more than the boss unless you have another job to step into.

Don't forget when a day seems long that thinking about it will make it seem longer.

APPENDIX.

CHAPTER XIV.

FAN SYSTEM FOR WOOL.

From J. U., Louisville.—I wish your magazine to give me a few ideas in regard to a fan system. I have had the pleasure of reading the articles by William H. Hayes, and I believe he is the man to ask. The fan in question is to convey wool from a dryer to a building about one hundred feet away.

The parties told me they wanted an 18-inch pipe run along the wall 5 feet from the floor of one building and go through a window to the other building 15 feet, and down to a wool dryer. I enclose a sketch of the plan. I asked what size fan was to be used. They told me they had not gotten the fan yet, but to go ahead and put up the pipe and they would get the fan. I put the pipe run along the wall 5 feet from the floor of 12 inches in diameter and an exhaust which was 11¼ by 10½ inches. When everything was in readiness they tried to force the fan to blow from the back of the hopper with the exhaust. The sequel was, it blew out of the hopper. I told them

they could not work it that way. I wanted them
to suction it from the hopper. I asked them if
the paddles would hurt the wool and they did not

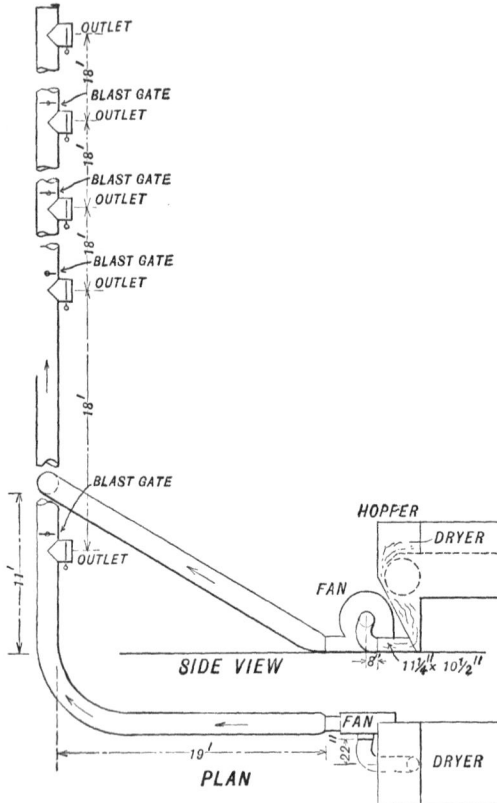

FIG. 34.—PLAN OF FAULTY BLOWER SYSTEM.

know, but on trying it they said it made the wool
lighter. One foreman then said it blew too hard
and tore the wool. The machinist then reduced

the speed of the blower to 800 revolutions per minute, which caused the pipe to choke up.

The question I want to ask is, is the pipe of too great area, 18-inch diameter, to correspond with that on the fan, which is 12 inches? Also, can the fan blow the wool through the pipe without suction (that is, going through the fan)? I have two elbows of 5-foot radius, with no chance to choke up.

I have put up pipes for exhaust systems for shavings and dryers, but I never had wool to contend with before. I would like to have Mr. Hayes give me some pointers on this. The building to which the wool has to be blown is about 15 feet from the main building, where the fan and dryers are, the pipe in this building being about 76 feet long and has five 18-inch drops, each being in front of a machine to which different grades of wool have to go. There are dampers to close and open, which are made on the order of a blast gate. The fan sets 8 inches from the hopper on the side, and 22 inches from center in front. The paddles are 24 inches, that is, 12 inches from center. I should like to know if it is advisable to blow the wool through the fan or suction it through it. We had a little difficulty with the wool catching on the fan shaft.

Answer.—I would advise J. U. to install a single 40-inch fan which has a 16-inch outlet. This will clear the 18-inch suction pipe. Order the fan with a cotton and wool wheel, a cut of

which is herewith shown in Fig. 36, and speed the
fan up to 1,200 revolutions per minute. As shown
in the side view, Fig. 34, the center of the fan
inlet is 8 inches from the outlet of the dryer.
This necessitates an elbow with but a 2-inch throat
radius. The radius of this elbow should be at
least 16 inches at the throat, which means that
the center of the fan should be 24 inches from the
outlet of the dryer.

The system as it is at present will never be
made to work successfully. A 12-inch inlet is en-

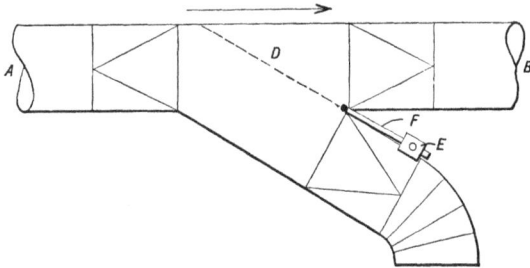

FIG. 35.—AN ANGLE GATE VALVE.

tirely too small to take care of an 18-inch pipe.
The area of an 18-inch pipe is practically 254½
inches, while that of a 12-inch pipe is only 113
inches. Furthermore, the right angle branches
and blast gate stops are a serious detriment to
the system if they are installed as shown in the
plan.

An angle gate valve, a sketch of which is here
given in Fig. 35, should be used at each of these
depositing outlets. This valve, which is con-

structed the same as a shavings switch, details
of which have been given, may be made at
an angle of 45 degrees or less. In the
cut A represents the inlet, B and C the
outlets. The dotted line at D represents the
gate. This gate must be solidly constructed and
made to fit the shell of the valve snugly. A counterbalance is shown at E, which must be so placed
on the bar F that the gate will lie evenly on the

FIG. 36.—WHEEL FOR COTTON AND WOOL.

bottom at all points, and yet can be raised to the
top with little effort.

With the use of this gate it will be seen that
the wool is not driven against a dead wall, also
that no consequent back pressure of air is produced.

Care must be taken to construct this work so
that all inside surfaces will be practically smooth,
as wool, cotton and other fibers will adhere very
quickly to projecting edges. The idea of blowing

the wool into the pipe as it falls to the bottom of the hopper is, of course, impractical.

A certain form of jet could be made, which would create a suction by using a strong blast, but the blast itself would tear the wool fibers to pieces, so there is no need to discuss it.

WANTS HELP ON A FAN SYSTEM.

From J. H. C., Maysville, Ky.—Enclosed find a rough sketch of a piping system, which was put in service about ten years ago. During the interval, the capacity of the plant has been doubled, and consequently the system has been giving poor results. I have been called upon to remedy this if possible. I should like, through your magazine, to obtain the assistance of Mr. Hayes.

Answer.—It is to be regretted that J. H. C. has not gone a little more into detail, as problems of this sort are always interesting and their solutions instructive. The writer has no way of knowing what sort of a plant he is contending with, whether a wagon shop, furniture factory or sash factory. Nor do we know what sort of a fan is in use. It seems to be a double, and is marked 32 inches, but this surely cannot be the diameter of the case, which is the dimension meant when fans are measured.

In view of the fact that an 18-inch inlet is shown, we take it for granted that the exhauster is a double 45-inch. Such a fan would have about a 28-inch diameter wheel. Again, we do not know whether the driving power is sufficient and what conditions prevail on the other side of the fan, or on the discharging end of the job.

These points are all of great importance and should be considered carefully, in order to solve the problem, or even to attempt a solution intelligently. With the information and sketch furnished by the correspondent we can only assume that the trouble lies in the suction pipe. This may well be the case, since the system as shown has in the rough sketch no semblance of proportion or proper arrangement.

There is a stretch of 14-inch main probably 40 feet long, with four 6-inch, two 8-inch and two 4-inch pipes tapped into it, and no tapering course. There is only one taper in the main and that was evidently put there in order to accommodate one 6-inch pipe on the end.

The dotted lines in the sketch show the main, as the writer has altered and proportioned it. (See Fig. 37.) The solid lines are the lines of the correspondent's sketch. I show the main running straight, thus eliminating two elbows. This should be done if possible. It may be impossible owing to location of shafts and belts, but as there is nothing shown to indicate this I assume that these two elbows were put into the main in order to run it down the center of the mill between the two lines of machines. I have repeatedly stated that elbows should be dispensed with wherever possible, es-

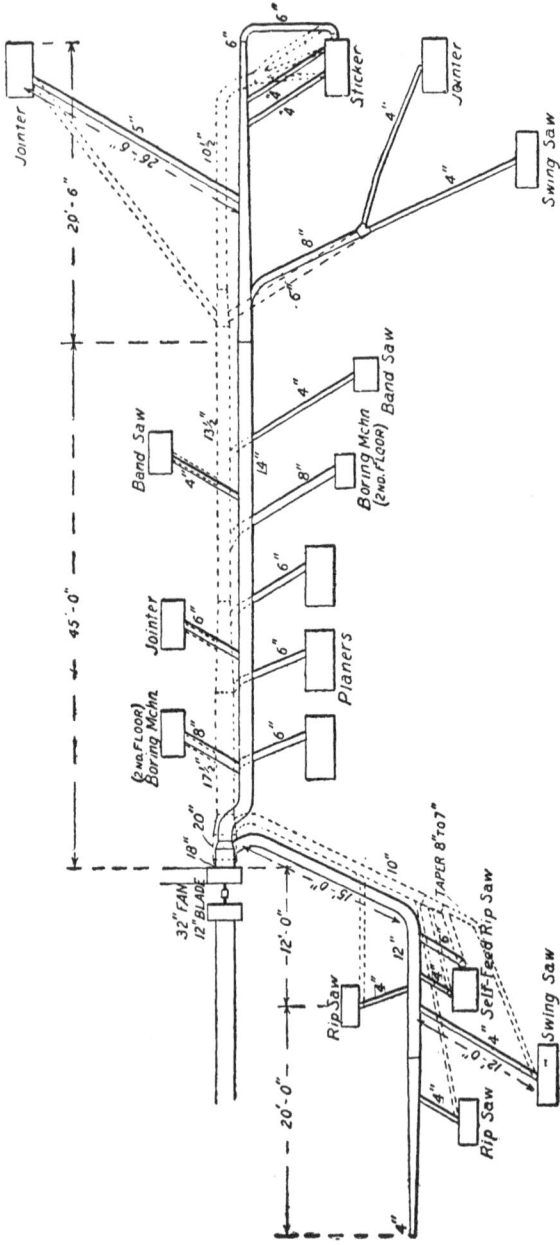

FIG. 37.—PLAN OF FAULTY SYSTEM SHOWING SUGGESTED CHANGES.

pecially in large pipes, on account of the large amount of friction they create.

The correspondent shows two boring machines, each on the second floor and each having an 8-inch pipe.

I do not know what sort of boring machines these are, but I doubt if both together would need so large a pipe as that shown. This is the biggest "snag" of the trouble, and the only way out of it is to eliminate these 8-inch pipes altogether, run a 6-inch floor sweep to each machine, and sweep up only when two of the lower planers are not running. No allowance in the main should be made for the 6-inch sweepers.

These planers should have "cut offs" in the pipe, but I would not advise "cut offs" in any of the other pipes. The reason for eliminating the two 8-inch pipes is that the main pipe is overloaded to an extent equal and a little in excess of the area of these two pipes.

Now, as to proportioning the main. The inlet of the fan is 18 inches. Enlarge this to 20 inches, as shown by using a short taper, then put on a "fan joint," a description of which has been given in a previous number. Tap in the 10-inch tee as shown in the butt of a taper from 20 inches to 17½ inches. Immediately beyond the 6-inch pipe from the planer nearest the fan make a taper in two joints, from 17½ inches to 13½ inches. Tap the next two planers and the jointer into that. As before stated, no allowance is made for the two sweepers. Tap the 4-inch from the band saw into the straight main, then taper from 13½ inches to 10½ inches, tapping the next two 6-inch pipes into the "butt" of the taper.

Run the 10½-inch pipe over to the sticker, using a three-way branch for the three pipes.

Direct the operators positively that this sticker shall always be left wide open. Indeed, all of the machines on that main, except enough to secure a suction for the sweepers when they are in use, should be left open.

I would be willing to bet a new hat or a bottle of "Old Crow," which is probably a better bet in Maysville, that the 14-inch main shown in the correspondent's sketch is half full of shavings at the present time, unless J. H. C. has recently cleaned it out.

The area of a 14-inch pipe is 154 square inches. The area of the pipes tapped into this 14-inch main before it is tapered is 213 square inches, 59 inches in excess of the area of the main pipe itself, and there are still four machines to be taken care of.

One would need to go far indeed to find a more remarkable instance of thoughtlessness or haphazard methods. The correspondent says that the suction is almost dead on the end. I could easily believe that it is dead and buried.

Again, I have changed the pipe running to the machines in lower left hand corner of the sketch from 12-inch to 10-inch, and even 10-inch is a trifle in excess of what the branches figure up. I have eliminated an elbow in this pipe, as will be seen, by changing the course of the branch pipes. As this alteration does not necessitate any elbows in the branches, the advantage of this procedure is obvious.

In altering the course of the big main from the fan it will also be seen that no extra elbows are needed in the branches, and the addition to their length is not enough to prove a detriment. If this main pipe cannot be run straight follow the old course, but taper it as shown.

A PROBLEM IN BLOWER WORK.

From H. F. B., Brooklyn, N. Y.—I am a subscriber and note what you say about inquiries to practical questions. I enclose sketches of a blowpipe job for a brass foundry on which I wish to ask your advice. The main pipe is to be 60 feet long and to have 16 inlets. Would you advise me to taper it gradually from end to end as shown in Fig. 38 or to taper it as in Fig. 39? The owner wishes to build a brick vault for the blower to discharge air and dust into. Do you think this will work? How large should the vault be, what should be the diameter of the outlet, and will the 14-inch blower be sufficient for the 16 inlets? I notice that W. H. Hayes speaks of such a vault in your March issue.

Answer.—I would not advise tapering the pipe gradually, since it is quite an expensive sort of pipe to make and is really not the best in point of efficiency. In all emery wheel and buffing work the main pipe should be straight on the bottom, as shown in the appended sketch. As a matter of fact this is not a bad way to taper the main suction for any blow-pipe job, though it is not a practice universally followed, because it is more expensive to lay out than the ordinary tapering joint.

It will be seen that there are no pockets on the bottom of a pipe so tapered. It must be remembered that the heavy material has a tendency to travel along the bottom and it is always the heavy material that will accumulate in a pocket.

The lint from a buffing wheel, which is continually tearing off and flying into the hood, becomes heavy and mean to handle when it accumulates in a pipe, as it often does, and becomes more or less coated with the dust it accumulates.

It is also a common occurrence for a nut or washer to fall into the pipe through inadvertence or carelessness on the part of the operator. To prevent this from reaching the fan a large tee joint called a trap is placed on the bottom of the pipe near the fan, as shown in Fig. 40. A cap is fitted to the end of this and it should be cleaned out occasionally. The throat of this tee, as shown at A, should extend into the pipe about one inch. Plenty of hand holes should be provided, as suggested by H. F. B.

As regards the vault, whether of brick or any

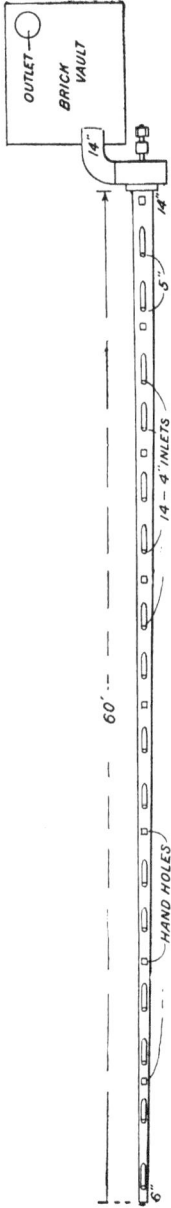

FIG. 38.—CORRESPONDENT'S SKETCH. (PLAN.)

FIG. 39.—ALTERNATE METHOD SUGGESTED BY CORRESPONDENT.

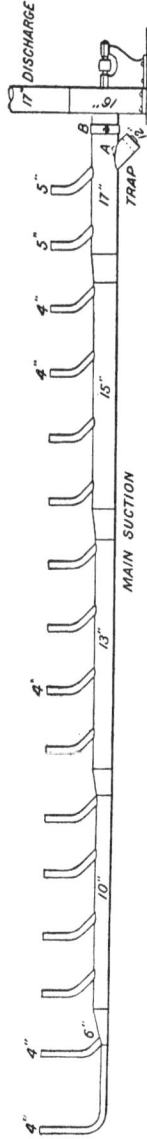

FIG. 40.—PROPER METHOD OF CONSTRUCTING SYSTEM. (ELEVATION.)

other material, I have never seen one that was
not a nuisance if the material is blown directly
into it. If the job in question is out in the country
the vault might be feasible, but I certainly would
not advise any one to try it in the city, especially
with the dust produced by the polishers blowing

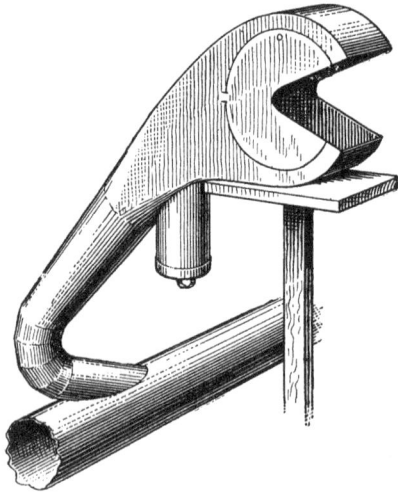

FIG. 41.—SUITABLE HOOD FOR EMERY WHEEL.

into it, for it is certain that 50 per cent. of that
dust will blow out of it and settle in the neighbor-
hood, unless some system of screens is used, and
this, I think, would involve an expense greater
than the cost of a dust separator. If a vault is
used it should be at least 12 by 12 by 15 feet and
have an outlet three times the area of its inlet.

It will be seen that the fan shown in the sketch

has a 16-inch inlet, while the large main pipe is 17 inches. It is the practice to figure the size of the main and then reduce one inch at the fan inlet, since the smaller fan creates less friction at the inlet and thus will have about the same efficiency as the next size above it, if the discharge pipe is also increased one inch.

WANTS SIZE OF SEPARATOR.

From O. R. & Sons, Jersey City.—I would like information as to the size of separator required for an exhaust system to remove the dust from buffers used in finishing brass. The outlet of the blower being used is 16 inches and the conditions are such that two right-angle elbows are required between the blower and the separator.

Answer.—In reply to the query of O. R. & Sons I send herewith a sketch, (Fig. 41A), with measurements and an explanatory list showing proportions. Naturally the size of a separator is determined by the area of the pipe or pipes that enter it.

The sizes of the several parts of the separator are worked out in proportion.

The rule here given is a safe one to follow in any and all cases. Thus the wind outlet is three and one-half times the area of the inlet; the diameter of shell is twice the diameter of wind outlet; the hight of the shell is one and one-half times the diameter of the inlet; the hight of cone twice that of shell, and so on.

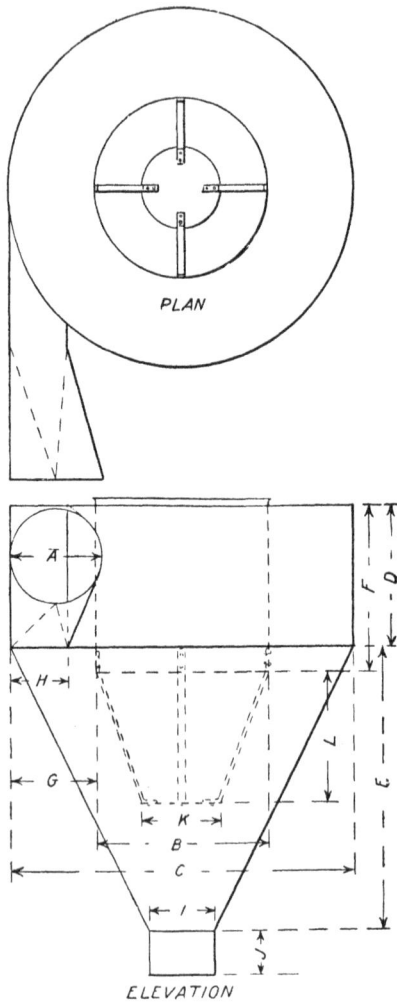

PLAN

ELEVATION

FIG. 41A.—PROPORTIONS OF SEPARATOR.

In the writer's opinion the wind outlet should in all cases be left free if possible. When a separator is working rain does no damage, unless grain dust or cement is being handled. When it is not in use for any length of time it should of course be covered up to prevent rusting, since the galvanizing is soon worn off by the action of the rapidly revolving material.

A separator of the size required in the case of O. R. & Sons may be made of the following gauges: Shell, No. 18 iron; inlet and transformer, No. 18 iron; cone, No. 20 iron; top, 1-inch pine lined with No. 24 iron above and below. This top should be soldered on the outside.

The disc or check draft shown is simply a round piece of metal, heavy enough to hold its shape, which is held in place by four bars of iron riveted to it and to the tubular guard as shown. Without this disc the material after reaching a point near the dust outlet would be caught by an updraft and carried out the top. Its location is a matter of little importance so long as it is on the perpendicular center line of the dust outlet.

I do not think it is necessary to solder a dust separator except on the top, if the top is made of wood. In my long experience in this line of work I have never soldered one or seen one soldered, and they were sold in those days for a price that would buy an automobile today and no runabout either.

The size of the dust outlet depends upon the

nature of the material to be handled and also the amount or volume of the same. If for long stringy shavings the collar should be very large as it will clog easily. The same applies to a small extent if a large amount of planer shavings are handled. In the present case a 10-inch outlet is large enough.

If it is deemed necessary to protect the material from rain use a canopy top and place it at least a distance equal to the diameter of the wind outlet above it, in order to give the escaping air free egress. Following are the proportion of parts for separator:

A, diameter inlet; B, wind outlet or tubular guard 3½ times area of "A"; C, diameter shell twice diameter of "B"; D, hight of shell 1½ times diameter "A"; E, hight of cone twice the hight of shell; F, hight of tubular guard ⅙ longer than "D"; G, expansion chamber ½ diameter "B"; H, width of transformer ⅔ of width of "G"; I, governed by circumstances; J, any length desired; K, disc for check draft 2 inches larger than "I"; L, location of check, about half-way between bottom of tubular guard and dust outlet—T.

W. H. H.

INQUIRY FOR CAPACITY OF BLOWER.

From O. R. & Sons, Jersey City.—Herewith we send you a photograph of separator constructed in accordance with the advice given us in your December number. The separator works to perfection, and we thank you for the reliable information, but it happens that the conditions have become such that we are in need of further information in regard to our exhaust system, which is employed to remove the dust from buffers used in finishing brass. As our buffers are between the sizes of 6 and 14 inches, we are now required by law to increase the size of our pipes 5 inches

FIG. 42.—SUCCESSFUL SEPARATOR.

in diameter. As we have seven machines with two pipes to each we have altogether fourteen pipes. The inlet to the blower is 16 inches in diameter. What total air capacity do we require and can this be made up in speeding the blower?

Answer.—We would recommend first that the correspondent try to get out of his trouble by speeding his fan. It is not likely that he may escape some expense by doing so. As regards the "air capacity" or area, the areas of the fourteen 5-inch pipes figure up 274.8 square inches and the area of the 16-inch fan inlet is 201.06 square inches.

If speeding his fan will not overcome the difficulty then he may use the same fan and enlarge the main suction directly it leaves the inlet to 17 inches in diameter and taper down as shown in the accompanying sketch. This will reduce the friction in the main to some extent and may be the only change necessary.

If still it does not work then enlarge the discharge pipe to 17 inches from the fan to the separator, changing the transformer on the separator to a 17-inch diameter round.

Personally we hardly think these changes necessary as the material handled is very light, easily handled and small as to quantity. It is always necessary in a job of this description to put hand holes at frequent intervals in the suction and cover them with slides made to fit reasonably tight.

The lint from the buffers or "rag wheels" often gets into the main suction and lies on the bottom, accumulating there until the suction is greatly reduced.

These hand holes give access to the main and enable the operator to clean out easily.

The purpose of the 12-inch trap shown in Fig. 43

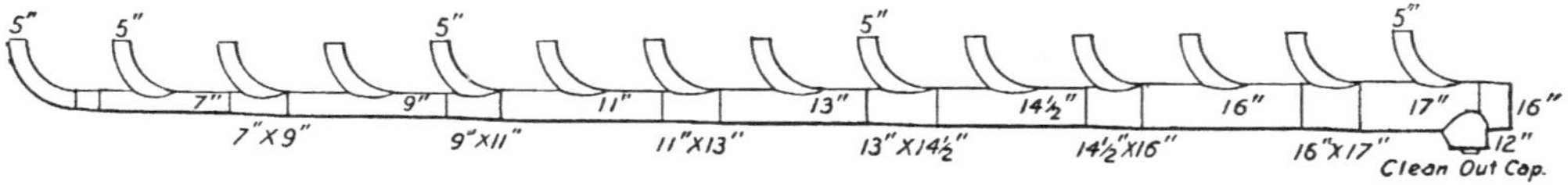

FIG. 43.—TAPER OF MAIN SUCTION.

is to catch washers and other pieces of
machinery which the operators often drop into
the hoods. A cup is fitted into this tee which
should be kept closed when the fan is running.
If the fan is set above the machines of course this
expedient is not necessary.

I will say in conclusion that the main suction
as shown in the sketch is not figured out "according to Hoyle," but is figured on the basis of using
a 17-inch pipe.

It will be seen that the area of a 17-inch pipe
is 229 square inches, while the actual area needed
is 274.8 square inches.

I have reasoned that perhaps by adhering to
the above layout the correspondent can work in
some of his old pipe in case it is found necessary
to make a change.   W. H. H.

POWER FOR EXHAUST FANS.

From M. F. H., Philadelphia.—Your readers
will be greatly indebted to Mr. Hayes for his recent article on the efficiency of the exhaust fan and
especially for the comparisons of power used by
fans of different sizes.

The data given, however, would be much more
valuable if your readers were told how to determine the exact saving in power by a careful se-
lection of any fan larger than that just necessary
to do the work; for instance, it is stated in the
small table that a 45-inch fan with a speed of 1,300
revolutions per minute will do a given amount of

work and use 11⅔ horse power. Will the 50-inch fan at 1,010 do the same work, and if so, how was the speed determined?

It is fair to assume that a 60-inch fan running at 650 revolutions per minute will do the work of a 45-inch one at 1,300 revolutions per minute, but few readers will know how the speed of 650 was figured.

With this point made clear, the tables given will be of very great help to those engaged in this class of work.

Answer.—In the February issue M. F. H. points out that the readers of "Practical Exhaust and Blow Piping" will be interested to know how slow speeds in exhaust fans are figured.

The figures given in the article in January (Chapter X, page 66, herewith) were taken from a fan catalog and were given to the reader for what they were worth.

These catalogs may be obtained from any blower manufacturer on application and they contain full information on speeds, power and volume.   W. H. H.

HOOD FOR CHEMICAL LABORATORY.

From L. J. D., St. Louis.—I would like to know the best material and gauge to use and also how to construct a hood for a chemical laboratory. The hood is to be 4 by 10 feet, with a 12-inch pipe leading from it to the ventilating flue. The hood is to carry fumes from sulphuric acid and other acids.

Answer.—We would recommend a hood as shown in the accompanying drawing. The hood is a double one, the combined measurement on

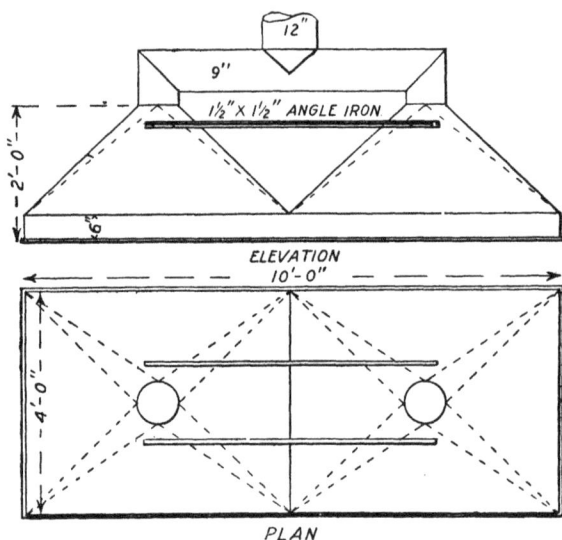

FIG. 44.—CONSTRUCTION RECOMMENDED FOR HOOD.

the bottom being 4 by 10 feet, with hight to suit conditions; the idea being to get the direct draught from the pipes as near the fumes as possible and yet maintain an easy pitch in the hoods. The sketch shows about the right hight in proportion to the length.

If this hood were made single with one 12-inch

pipe leading from its center it would necessarily be too high to be effective, unless it could be placed very close to the object it serves.

The outlets are each 9-inch and may be run together as shown and submerged into one 12-inch pipe.

I would, however, recommend the use of four or five piece elbows if there is enough head room to permit it.

The hood should be made of No. 12 B. W. G. copper, thickly coated inside with lead. The pipes may be made of lighter copper treated the same way. Two angle iron braces are shown riveted to and connecting each hood. All rivets used should be copper coated with lead.

The pipes and indeed the hood could be made of galvanized iron and lined inside with sheet lead, but this operation would be more expensive than the copper construction, especially in the pipes, and the fumes that may escape from the hood would ultimately destroy it from the outside.

The hood consists of two parts, each made square to round, riveted together in the center, with a 6-inch perpendicular rim around the lower end stiffened by laying in a ⅜-inch round iron bar.

There should be a hanger on each corner and two in the center if it is to be hung from the ceiling.

I will say in conclusion that if the ventilating flue that is to carry off the fumes is of galvanized

iron, block iron, or tin, the action of the fumes
will soon destroy it.   W. H. H.

From J. F. S. Co., Chicago.—Recently we installed a blow-pipe system similar to the one described by Mr. Hayes in the July number of your
magazine (Chapter V, herewith). We wish to
ask him to give the dimensions of the separator
he would advise using on this system, as the one
we have installed gives us two inches of back
pressure at the intake. The dimensions of the
separator used on this job are as follows: Body,
5 feet; cone, 8 feet and diameter, 7 feet; air outlet,
46 inches, and dust outlet is 12 inches. Our
fan is running 925 R. P. M., and the material, instead of dropping rapidly, as in our other systems, keeps whirling around and drops gradually.
This factory uses oak, pine and bass wood. We
would appreciate an early reply, as we intend to
make changes.

Answer.—In reply to the query of J. F. S. Co.
will say that it is difficult to give proper dimensions of the separator he needs, in view of the
fact that he does not give the size of the discharge
pipe, the size of his fan, or the number and description of the machines he is piping. Enlight-
enment on any one of these conditions would
simplify the proposition.

He states that the fan is running 925 R. P. M.
He may be using a fan from one to three sizes

larger than the combined areas of his branch
pipes call for, in which case that information is of
no value. He may be, and doubtless is, using a
single 55-inch fan, in which case there is nothing
radically wrong in the proportion of the separator he describes if it is properly constructed. The
diameter of the shell is a little too narrow for
the size of the wind outlet if the discharge pipe
that enters it is 23 or 24 inches, which would be
proportionate. But the length of the perpendicular shell being 5 feet should in a measure offset this objection and give room enough for air
expansion.

The tubular guard may be too long or there
may be something inside the machine in the way
of baffles or deflectors which impede the progress
of the air. Again, the transformer or transition
piece where the discharge enters the separator
or the inlet itself may be too small. It should
be somewhat larger in area than the round pipe.
There may be a canopy top or some other description of cover over the pure air outlet or tubular
guard, which is set too close to it or working over
it automatically, either of which conditions is
wrong and should be rectified. The instant the
air leaves the round discharge pipe it should begin to expand. The object of a rectangular inlet
entering a separator is to form a narrow flow of
material as it strikes the wall of the separator,
thus the material is thrown to the outside and
kept there by centrifugal force. This should be

done without changing the course of the air too abruptly or choking it down in the least.

As I said before, it should be expanded rather than contracted. Once inside the separator everything should be as free as it is possible to make it, and have the machine perform its function. In the cyclone pattern of separator, a description of which has appeared, absolutely no baffles, deflectors or other obstruction should be used. There was never a separator made that worked properly that did not lower the efficiency of a fan to some extent. This is a well-known fact, and any statement to the contrary may be set down as pure "bunk." So I would advise the correspondent to look thoroughly for either causes before he lays the blame for the trouble to his separator and goes to the expense of making a new one.

Too much stress cannot be laid upon the fact that the air outlet should be absolutely free. The air enters the separator at a certain pressure. Once in the separator, though still in rapid action, the pressure is greatly reduced. Thus to be able to leave the machine as fast as it enters it the avenue of egress should be enlarged as near in proportion to the reduction of pressure as it is possible to make it and still confine the material. Therefore anything that is placed either inside the machine or outside over the air escape that impedes its progress in the least is simply making it that much harder for the fan to deliver the goods. To illustrate this take a can that will hold two gallons and put a ¼-inch nipple in each end of it and blow through it. You will feel as if the can were taking your breath away. Then cut a large hole in the can and repeat the performance; you'll find it a pleasure to blow in it if you have nothing else to do.

The correspondent speaks of the material "whirling around" and "dropping gradually" out of the separator. The speed of the fan has little to do with this. If no feeder is used I think this an advantage. It is certainly not a fault if the material eventually gets out. If there is so much handled that there is danger of clogging put a little spiral piece at the bottom of the cone extending from a point about 2 feet above the collar well down into the collar.

If the air is needed for feeders make this spiral longer; indeed, continue up until proper results are obtained. Two inches will be wide enough for it.

In conclusion, I would refer the correspondent to the article in the December number, entitled "Construction of a Separator," (Chapter IX, herewith). He will find there the method of obtaining measurements and proportions for any dust separator.   W. H. H.

FURTHER DISCUSSION OF SEPARATOR PROBLEM.

From J. F. S. Co., Chicago.—Referring to your Mr. Hayes' letter of explanation of our separator problem in your January number, will say that the type of fan we are using is a Double Garden City Cycloidal 50-inch slow speed fan. It is loaded to within 24 inches of its suction area on both intakes, and all machines piped are small, the largest being a 3-drum sander. The following is a list of the machines: Three shapers, one sticker, two planers 24 inches, two jointers, one sander as stated above, one dovetail machine, one band saw, six saw tables, one trim saw. First floor, main room, one saw table, one shaper, four belt sanders. Second floor, small room, six saw tables, one jointer.

The separator, as stated before, is 7 feet in diameter, the body 5 feet high, the cone 8 feet high, the dust outlet 12 inches. We have no canopy or elbow on the outlet of separator. The size of the transformer is 15 inches by 38 inches, which you will note is rather small, but before the round pipe connects with the transformer, we have put in a 6-inch air feed for the boiler, which will offset this. This separator is patterned after the one as described by Mr. Hayes in one of your issues.

The fan man made a mistake when giving us the speed of his machine, hence the poor suction. We know this fan will do the work at 1,100 R. P. M., but the proprietor is an obstinate person and refuses to run it any faster.

Also wish you would ask Mr. Hayes if he has had any experience collecting chalk dust; also the

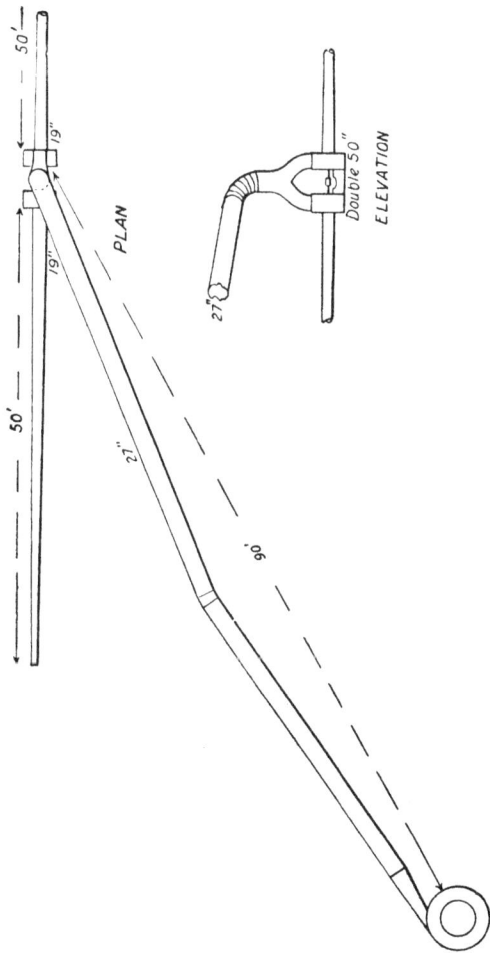

50'

19"

PLAN

19"

50'

27"

90'

27"

Double 50"

ELEVATION

FIG. 45.—SKETCH SENT BY J. F. S. CO.

least amount of back pressure that a collector makes when working under full load.

Answer.—In reply to this query will say that the correspondent's trouble, in my judgment, is due to the fact which he states, viz.: that his fan is not running fast enough. His discharge pipe is not excessive in length, and if his suction mains are properly proportioned there is nothing wrong with the system.

It would seem that the proprietor is standing in his own light in refusing to speed up his fan.

As regards the least amount of "back pressure" a separator will cause, will say that a properly constructed machine will not cause any.

What is generally termed back pressure is a loss of pressure due to friction, and in a separator is caused by transforming and by the change in direction of the flow of air. This, of course, would be more noticeable if the separator were placed at the end of a long distance pipe than if it were placed at the fan itself. In the latter case the loss in efficiency should not be more than 3 per cent.

I have had no experience with chalk dust, but would give it as my opinion that in its dry state it can hardly be successfully handled with an ordinary separator. W. H. H.

WANTS TO SEPARATE DUST FROM SHAVINGS.

From J. W. S., Kansas City.—Will your magazine or its readers give me information

on how to separate sawdust from shavings after
it has been carried into the separator, which we
are now building in accordance with directions
furnished in your previous issues?

Answer.—We know of no way to accomplish
this. It is usually hard to tell the difference between coarse sawdust and dry wood shavings
after the fan is done with them, and even if this
were not the case there is no way of separating
the two after they have been run through the
same fan. Buy a small fan for your sawdust and
build a separator for it.   W. H. H.

SEPARATING SAWDUST FROM SHAVINGS.

From M. F. H., Philadelphia.—I note a statement in your journal, made in reply to the
inquiry of J. W. S., in which he is told that
there is no way of separating sawdust from shavings after both have passed through the fan.
This is being done in a number of the mills of this
city. We are now installing a new furnace feeder
in a box factory where the separator successfully
separates the dust from the shavings.

From N. B. P. & M. Co., New Orleans.—In
your April number we find an article under the
heading "Wants to Separate Dust from Shavings," in which the inquirer is told that it cannot
be done. For your information we desire to say
that we have a number of such separators in
operation and the separation can be done very
nicely.

DUST SEPARATOR FOR CARPET CLEANING PLANT.

From M. F. H., Philadelphia.—In reply to J. D. L., of Toronto,* I would advise him not to use a separator on a dust exhausting system in a carpet cleaning plant. He will find if he does, that both dust and air will blow clear through the

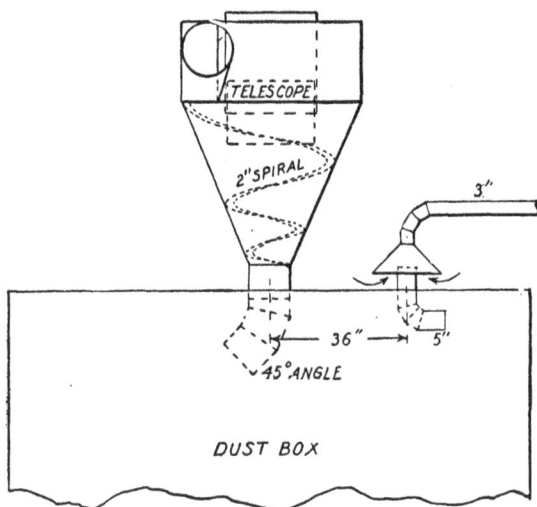

FIG. 46.—DUST SEPARATOR FOR CARPET CLEANING PLANT.

separator and little if any will fall into the dust box. The dust appears to be so intimately mixed with the air that it will not separate.

From W. H. H.—If J. D. L., Toronto, contemplates using a "Cyclone" separator such as has

*J. D. L. inquired if dust separators are in use in carpet cleaning plants.

been described in your magazine, we would suggest such additional features as are shown in Fig. 46. The dust from the carpet cleaner will be very fine and not easily controlled.

Make the tubular guard according to directions previously given as to length and diameter, and put an extension in it that can be raised and lowered. Construct a spiral in the cone as shown in the sketch, going at least two and one-half times around it. This spiral should not be more than two inches wide.

When operating the system lower this tubular guard as far as is possible without interfering with the suction pressure at the fan, and fasten it when the proper position has been determined. The inside sleeve can be slotted in about six places, equally spaced around the circumference, the slots extending perpendicularly, and can be secured to the outer sleeve with bolts, nuts and washers.

Set the separator at about the center of the dust box, as shown, attaching a 45-degree angle to the collar, so placed as to cause the angle to fall in the opposite direction from the 5-inch vent pipe. Set the 5-inch vent pipe about 36 inches away from the center of the dust outlet and put an elbow on it inside of the box pointing directly away from the 45-degree angle. This pipe is to supply ventilation without allowing the dust to escape from the box which must be made absolutely tight and kept perfectly dry.

If the material becomes wet while in the separator it will soon cake and accumulate. The elbow on top of the separator should be made to work on a swivel and swing with the wind, as shown in

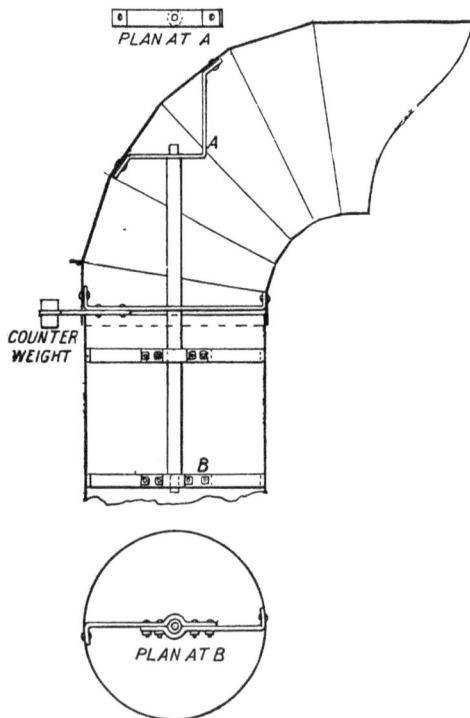

FIG. 47.—SECTIONAL VIEW OF SWINGING ELBOW.

Fig. 47, which shows a cheap and effective way to construct the swivel elbow.

This separator should be soldered on top and puttied inside along the laps.

Inasmuch as we do not know the size of the fan

we cannot give the size of separator nor the measurements of any of the work pertaining to it,
but we suggest that the dust box cannot be made
too large. The size has nothing to do with the

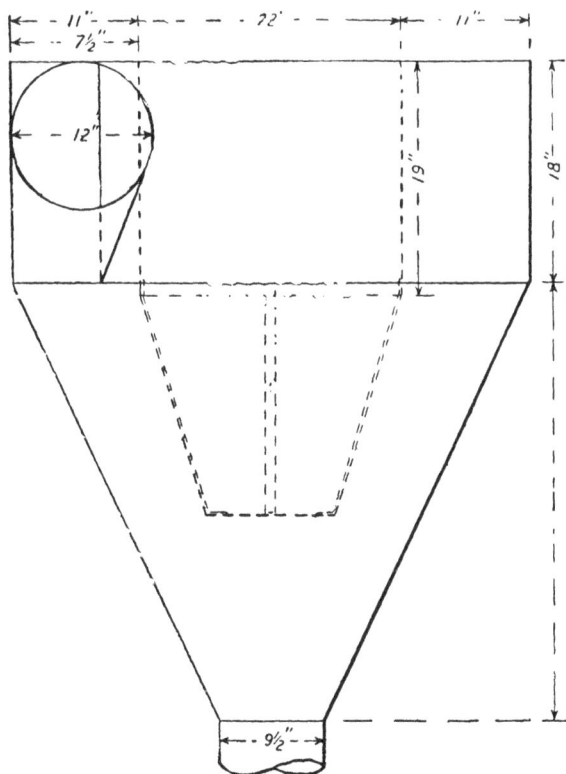

FIG. 48.—DIMENSIONS OF SEPARATOR.

quantity of dust it is to hold; the dust should
settle as far as possible from the vent of the box.
The cone shown over the 5-inch vent pipe should

be about 12 inches in diameter and 6 inches high,
and set near enough to the vent to catch the dust
without forming a strong draft in the box.

We know of no cases where the centrifugal
separator is being used for this particular purpose.

SEPARATOR FOR SINGLE PLANING MACHINE.

From S. F. P. Co., Warren.—We wish to construct a separator drum to handle shavings from
one planing machine. This will be operated by a
30-inch fan running 800 revolutions. Will you
give us the proper dimensions for the separator?

Answer.—Our correspondent does not give the
size of the planing machine but says he will use a
30-inch fan at 800 revolutions. If he is using the
centrifugal type of fan such as that sold by the
American Blower Company or the B. F. Sturte-
vant Company, he will not be able to handle shavings with it at a speed of less than 2,000 R. P. M.
and as a matter of fact it should run 2,500 R. P.
M. However, we present in the sketch herewith
given the dimensions for a separator to accommodate a 30-inch single fan. (See Fig. 48.)

W. H. H.

EXHAUSTING DUST FROM FOOD MACHINES.

From M. F. M. Co., York, Pa.—We wish to exhaust the dust from five machines, sketches of
which, containing measurements, we send herewith. The dust is of the kind that passes off of
material such as corn fodder, corn, oats, etc. The

hoods for three of the machines will be similar in size to that shown in the smaller sketch, which will stand from 6 to 12 inches above the tops of the machines, while the other two, shown by the

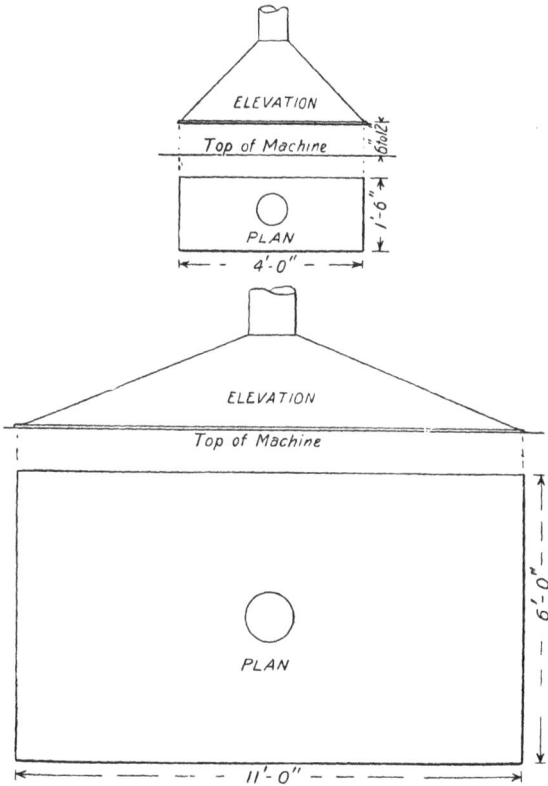

FIG. 49.—PLAN OF MACHINES.

other sketch, will rest on the tops of the machines, being thus closed all around.

We wish to know what size of pipes to use leading off from these hoods, the size of blower, and the speed necessary to create plenty of suction. The blower can be placed above or below the machines, as we have plenty of room. The space occupied by this machinery is about 20 feet square, and we will not require too long a run. The dust collector will be placed about 15 feet above the machine.

Answer.—In replying to query of M. F. M. Co. will say that if the dust merely rises from these machines without any help from the action of the machine itself then the fan must lift it.

If it is very light the larger hood, being fastened to the machine, will be effective if constructed as shown in the subjoined sketch. This is practically a double hood with a rim around the bottom, and each part transforming to a pipe 10 inches in diameter. Unless a sufficient quantity of air can be taken in through the machine itself, inlets should be cut near the bottom of the hood to allow as much air to enter as the 10-inch pipes will handle. It must be understood that air should enter the hood at all points if possible, else the material will drop and pile up where there is no air current.

The smaller hood setting 6 inches or one foot away from the machine will have little effect unless the material is thrown toward it. If it sets 12 inches away from the machine there will be 5½ square feet of area. It will be seen that it would take an enormous pipe and consequently a

very large fan to move that amount of air at a
sufficient velocity to be effective. I would suggest that as much of that space be enclosed as
possible and the hood constructed in the same
manner as the large one, letting the sides of the

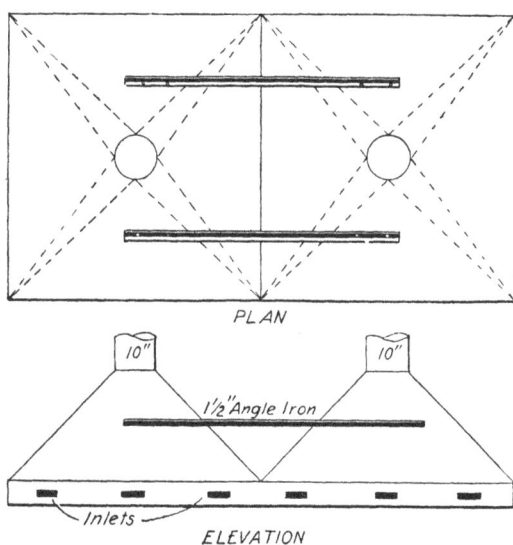

PLAN

ELEVATION

FIG. 50.—HOOD FOR FOOD MACHINES.

transformers approach the outlet at an angle of
45 degrees on the widest ends, and making the
outlets 6 inches in diameter instead of 10 inches.
Then select a single exhaust fan having a 17-inch
inlet and run it up to catalog speed.   W. H. H.

SIZE OF DUST COLLECTOR.

From L. G. F., New York.—Will you kindly give
me the size of dust collector for the accompany-

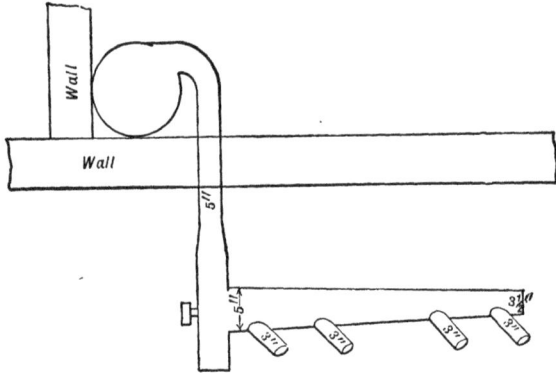

FIG. 51.—DIMENSIONS OF DUST COLLECTOR WANTED.

ing sketch, and also the width of spiral? The
system is to be attached to buffing wheels.

Answer.—Replying to the inquiry of L. G. F.,
who desired the proper size for the dust collector
used in job shown in Fig. 51, in which the system
is attached to buffing wheels, will say that Fig. 52
is a sketch of the form of collector required.

A is a tubular guard three times the area of C.

B, the width of shell is twice the diameter of A.

C is the diameter of discharge from the fan.

D, the width of inlet entering shell, should be
two-thirds the width of C.

E, the hight of shell should be two-thirds
greater than the diameter of C.

F, the hight of the tubular guard is 2 inches
longer than E.

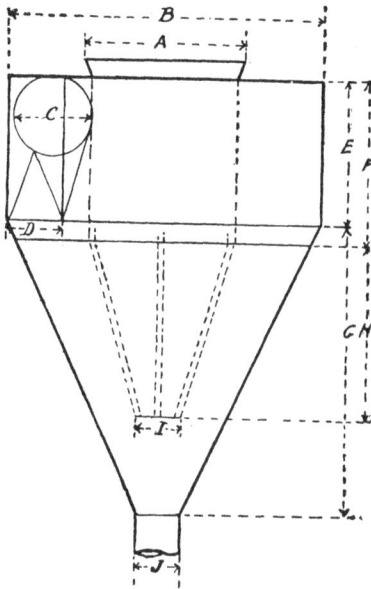

FIG. 52.—FORM OF COLLECTOR REQUIRED.

G, the hight of the cone is twice the length of E.

H extends down to where the diameter of the cone is twice the width of I.

I is a metal disc fastened with bar.

J the diameter of the collar is governed by the amount of material handled.   W. H. H.

A CENTRIFUGAL EXHAUST HEAD.

From G. T., Cincinnati.—Will you please publish a detailed drawing for an 8-inch steam condenser for the top of an exhaust pipe sometimes called exhaust head? I believe some are made to

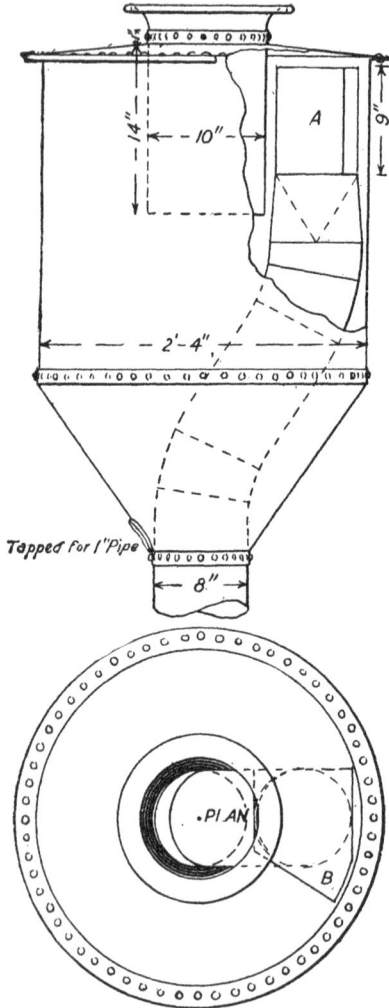

FIG. 53.—DETAIL OF CENTRIFUGAL EXHAUST HEAD.

work by centrifugal force, and am told that this type is the best.

Answer.—We present in Fig. 53 a detail of a centrifugal exhaust head. The contrivance as shown is simple and effective. It will be seen that the top of the escape pipe at A is transformed to a square cornered elbow. This is necessary, since the space is too small for a round elbow. As shown at B in plan the top elbow is brought around on the outside, so that the steam will not be driven against the side of the shell too abruptly. A one-inch drain is shown at the bottom of the cone. This may be either a malleable flange or a piece of half-inch steel plate tapped for one-inch pipe and riveted to the. cone, also well soldered. This whole machine should be thoroughly sweated with good half and half solder.

REMEDYING DEFECTIVE BLOWER SYSTEM.

From J. D. & Son, Chicago.—We have been called upon to remedy a defective blower system. On examination we found that the contractor who

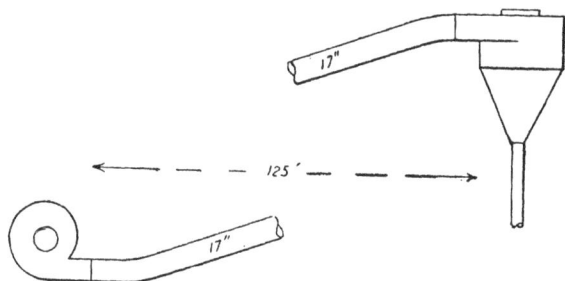

FIG. 54.—DEFECTIVE BLOWER SYSTEM.

did the job used an old discharge pipe, which had before been attached to a smaller fan.

The plant was recently enlarged somewhat, and a larger fan was installed, but, as we said, the discharge pipe was not changed.

The enclosed sketch shows conditions as they are.

It will be seen that the pipe is 17 inches in diameter. It should be 20 inches.

The separator at the end was made for a 17-inch diameter pipe and is 125 feet from fan.

We have to take this job on a contract and have competition, so would like to know the cheapest way to get the firm out of this trouble. How much back pressure is caused by this 17-inch pipe?

Answer.—There is one quite obvious way of relieving this fan, and that is to construct the 20-inch pipe, but it is not the cheapest way.

In Fig. 55 is shown an 11-inch tee tapped into

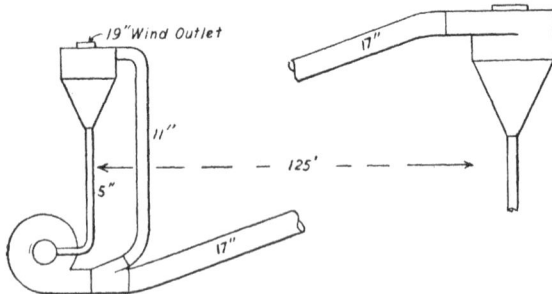

FIG. 55.—TEE TAPPED INTO FAN INLET.

the fan inlet; this should be tapped in as near the fan as possible. Erect a separator to accommodate an 11-inch pipe; a 5-inch dust outlet is large enough for the separator. Set this separator in some convenient place, as near the fan as possible, being sure that the dust outlet is high enough above the center of the fan to give the material a pitch.

Connect the separator to the discharge pipe with an 11-inch pipe.

The combined areas of the 17- and 11-inch pipes exceed that of a 20-inch pipe by a fraction.

There is no back pressure caused by the 17-inch pipe; the fan is simply taking in what air it can force through that pipe at a certain velocity. This being the case, the suction is weak, as the suction pipe is proportioned for a 20-inch discharge. If the areas of all the branches equaled, not exceeded, a 17-inch diameter pipe, there would be nothing wrong with the system.   W. H. H.

CONSTRUCTING HOOD FOR EMERY WHEEL.

From W. J. W., Canton, O.—I wish to ask your assistance in regard to some hood work, of which I enclose a sketch.

A in the sketch is a 20-inch emery wheel; B is the bottom where the grinding is done, and C is the direction in which the dust and emery fly. A hanger rod is fastened to the joists at E by screw eyes, which consists of a flat rod with a hole at F so that rod I can turn. G is a grinding buck of which H H are the handles. The casting to be ground is laid on G and the grinder takes hold of the handles and lifts up the buck so that the

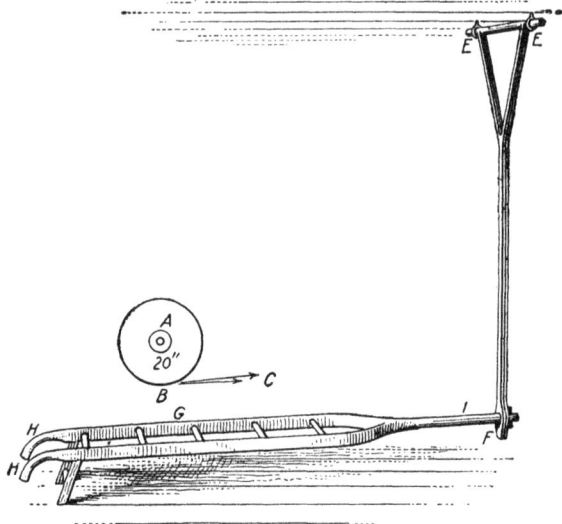

FIG. 56.—REPRODUCTION OF CORRESPONDENT'S
SKETCH.

casting will touch the emery wheel. The hanger
is 6 feet from the wheel, but can be changed if
desired. I would like to know how to construct a
hood to catch the dust from the wheel. I have 12
wheels to connect up. The main line of pipe to
the fan is 12 inches. There are two fans, and
two lines of pipe, one line on each side of the
room, and the main lines are up at the ceiling.
The fans have 12-inch inlets and 12-inch outlets,
with a 5-horsepower motor for each fan. I have
tried various hoods, but without complete success.

Answer.—In reply to the query of W. J. W.,
will say that the style of hood shown in Fig. 57

would be efficient in this case. It should not be
made any larger in proportion to the size of the

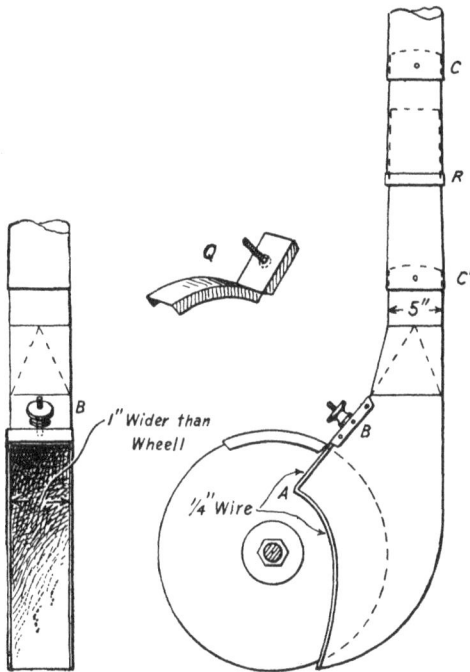

FIG. 57.—DESIGN OF HOOD FOR WHEEL SHOWN
· IN FIG. 56.

wheel than is shown, but should be made about 1
inch wider than the wheel. The point A should
extend down to a point about half way between
the rim of the wheel and the hub and a slide, shown
in detail at C, should be put in to take up the

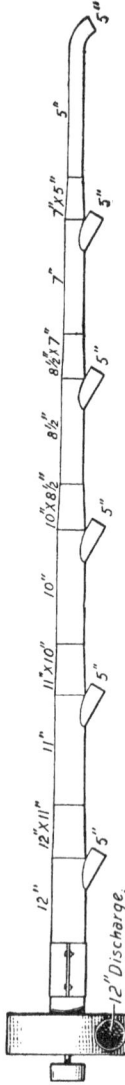

FIG. 58.—PLAN OF SYSTEM, SHOWING JUNCTION OF BRANCHES.

space as the wheel becomes smaller from wear. This keeps the suction near the wheel at all times.

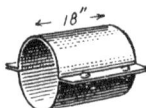

FIG. 59.—FAN JOINT.

A flanged piece B is riveted in; indeed, it is a part of the hood, and a slot is cut in its face. The piece shown at Q above, is made of the same gauge as the hood and narrow enough to slide under the flanged piece B. A bolt is fastened to Q as shown and after the piece is put in, it is held in position by a thumb nut.

This hood must be so fastened to the frame of the machine that it can be raised up or lowered down as the wheel wears and its lower end cannot come below the bottom of the wheel at any time. It should also be fitted with a slip and swing joint as shown in Fig. 57, so that it can be pulled in toward the hub and secured there or raised as the case may require. The correspondent does not state what sort of castings are ground on these wheels, but we will say in passing that if the castings are concaved and the concaved side is placed toward the wheel, of course this hood would be in the way, an if the hood were raised to suit such a condition and then a flat casting were ground, the dust from the flat casting would escape the hood. This is a problem which the man on the job must figure out for himself or else state very specifically what all of the conditions are.

The opening in this hood should be stiffened with ¼-inch rod. The hood should be made of No. 18 galvanized iron and its outlet should be 5 inches in diameter. As shown in Fig. 58 a 12-inch suction main figures just right for six wheels. The correspondent states that he has twelve wheels, six on each side of the shop. If he is using a single 30-inch exhaust fan, which has a 12-inch diameter inlet, its speed should be 2,350 revolutions per minute. The 5-inch tees should be tapped in just ahead of the taper butts as shown in Fig. 58.

A fan joint, shown in Fig. 59, should be put in near the fan so that the fan may be repaired without disturbing the pipe. All branch pipes should be 5 inches in diameter and tees flanged and riveted to the main. There should be hand holes on the top of the main at intervals of 15 feet. The telescope and swing joint is shown in Fig. 57. The joints C and C¹ are swings, the outside pipe being expanded and the two riveted together with two rivets directly opposite each other. Burrs or washers should be put on the "driven" head of these rivets. The outside or sleeve joint of the telescope should be reinforced by a piece of ⅛-inch by 1-inch band iron, shown at R.

W. H. H.

SPECIFICATIONS FOR DUST SEPARATOR.

From L. J. D., St. Louis.—Please give me information on the proper construction of a dust separator, and state if they can be bought complete and about what the price would be. The pipe from the fan will be 18 inches diameter.

Answer.—In replying to the query of L. J. D., we append a sketch giving size and proportions of dust arrester suitable for an 18-inch pipe.

The correspondent asks if they can be bought, made complete, and what the price would be. Replying to this would say, we can hardly give him an accurate estimate on the cost of one, but approximate it at about $50. We would suggest that the correspondent make it himself, if he is in the business; the cost would then be about $30. The shell and tubular guard of this machine should be made of No. 18 iron and the cone should be of No. 20, securely riveted and puttied inside after it is set up. I have before recommended that the tops of these be made of wood—1-inch pine sawed out in segments and lined above and below with No. 24 galvanized iron. If the correspondent has not the facilities for sawing this wood, he can make a top of No. 18 galvanized iron, pitching it enough to stiffen it. All of this work should be riveted, the rivets pitched about 2½ inches. The disc D, shown in Fig. 60, should be made of No. 14 iron and secured to the tubular guard A with ¼ x 1-inch bar iron, as shown.

W. H. H.

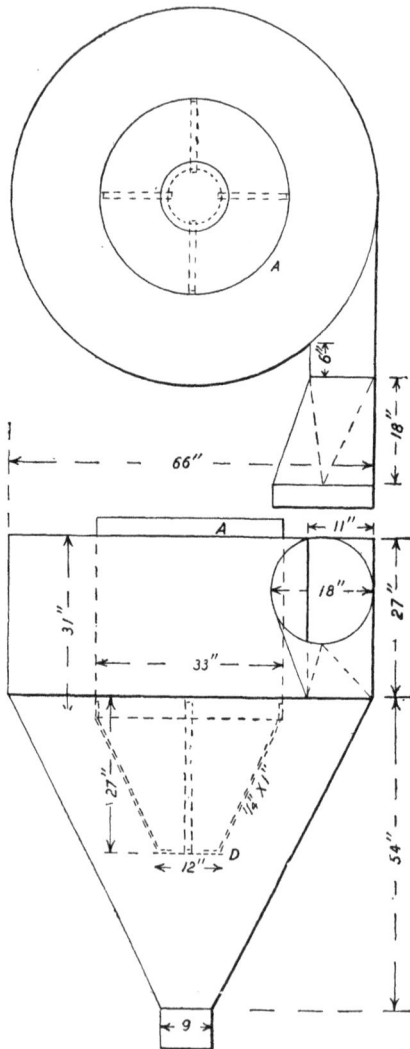

FIG. 60.—SIZE AND PROPORTIONS OF DUST ARRESTER.

CONNECTING SEVERAL FANS TO ONE SEPARATOR.

From W. C. Co., New York.—Is it practicable to connect three or four fans to one dust separator? I am figuring on a contract to erect a blower system for a man who owns a large business block, and who runs a wood working establishment on the ground floor. There are in this building three small wood working concerns who also want their refuse taken care of. The owner wants me to erect a separator large enough to take care of all these concerns, and one other that will come in later, and I am to erect feeders for the furnace, which of course is on the ground floor. This building is five stories high, and the separator is to set on the roof. Can I feed four boilers setting side by side with one separator?

Answer.—It would be practical enough to connect any number of fans into one separator if all of them were connected directly to the separator and most of them kept running. Fig. 61 shows how they should be connected.

Three pipes are shown connected. If more are to be used the shell should be made higher. They should enter the separator as shown, one above the other. It is not a good idea to put on another inlet, as shown by the dotted lines at A in plan, as the two volumes of air would tend to counteract and throw the shavings too far toward the center.

The pipes shown entering in the elevation are numbered 1, 2 and 3. If any two of these, or in

other words, if two-thirds of the amount of air
that the separator is designed for were taken from
it, the air current still entering it would be too
weak to cause enough centrifugal force to keep
the dust away from the tubular guard.

If you take this contract under a guarantee we
would advise you to insert a clause to the effect
that the separator will work properly with two-
thirds of the power in operation. If, as you state,
the separator is to be set on the top of a five-story

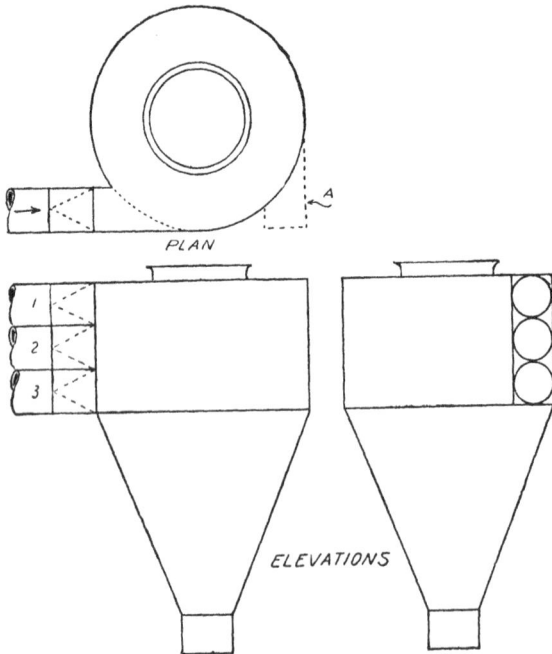

FIG. 61.—PIPE CONNECTIONS TO SEPARATOR.

building you could feed four boilers with it effectively. It is all a question of gravity.

W. H. H.

CLEAN AIR SUPPLY TO FEEDER PIPE.

From K. B., Los Angeles.—How can I get clean air into a feeder pipe? I recently erected a job on which there is a single feeder and no shavings bin.

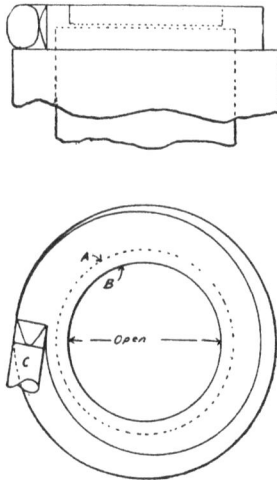

FIG. 62.—FORM OF HOOD RECOMMENDED.

The shavings, when not deposited into the fire, are dropped in front of the furnace, where the fireman can shovel them in. Blowing air in from the cone of the separator is all right so long as the feeder is operating, but when the shavings are being dropped on the floor the wind scatters them over the boiler roof. Can this be remedied?

Answer.—We recommend the correspondent to build a hood as shown in plan and elevation in Fig. 62.

"A" in plan is the wind outlet of separator. "B" in plan is the inside rim of the hood which skims around the outside of the wind outlet and gathers air into the hood. Be sure to have the outlet "C" pointing in the direction in which the wind is rotating.

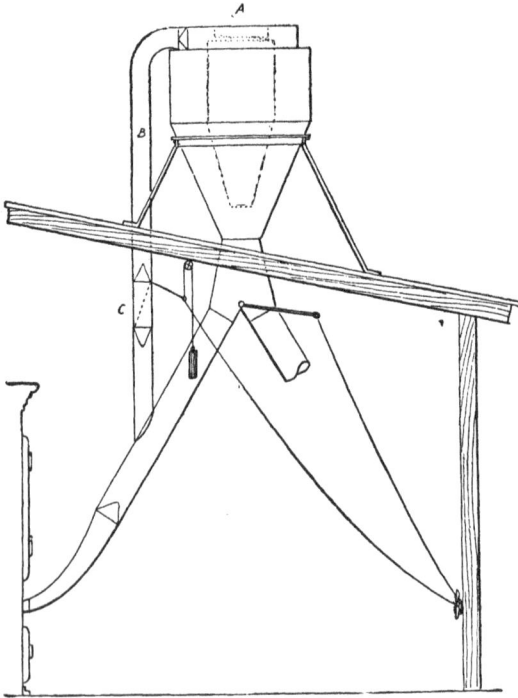

FIG. 63.—PIPE CONNECTIONS TO HOOD.

For a single feeder a 6-inch pipe should be large enough, which would bring this hood to a height of 7 inches. The inside rim should extend down about even with the top of the wind outlet, or a little above it, as shown at "A" in Fig. 63.

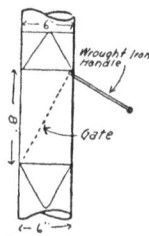

FIG. 64.—DETAIL OF VALVE.

The 6-inch pipe should be topped in above the feeder hinge as shown and the valve C, a detail of which is shown in Fig. 64, should be put below the roof and regulated with a cord and weight as shown.

When the feeder is down this valve can be closed, and as there is very little light dust caught in this hood it will not clog the pipe unless it is kept closed for an indefinite period.

This valve also regulates the flow of air.

W. H. H.

ALTERATIONS FOR INEFFICIENT EXHAUST SYSTEM.

From B. R. F., Goshen, Ind.—Herewith I send you a plan of blower system which is not satisfactory. The fan is 70 inches and speed is 750 revolu-

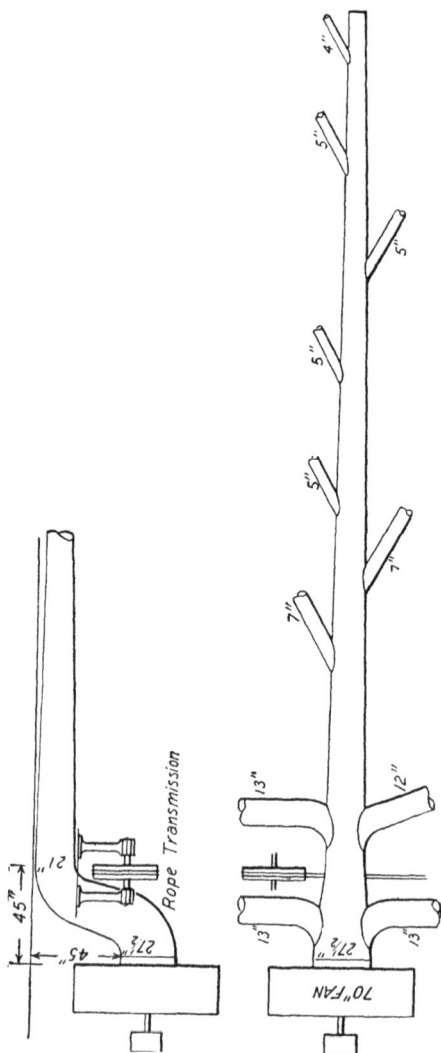

FIG. 65.—CORRESPONDENTS' SKETCHES SHOWING PLANS AND ELEVATION OF SYSTEM.

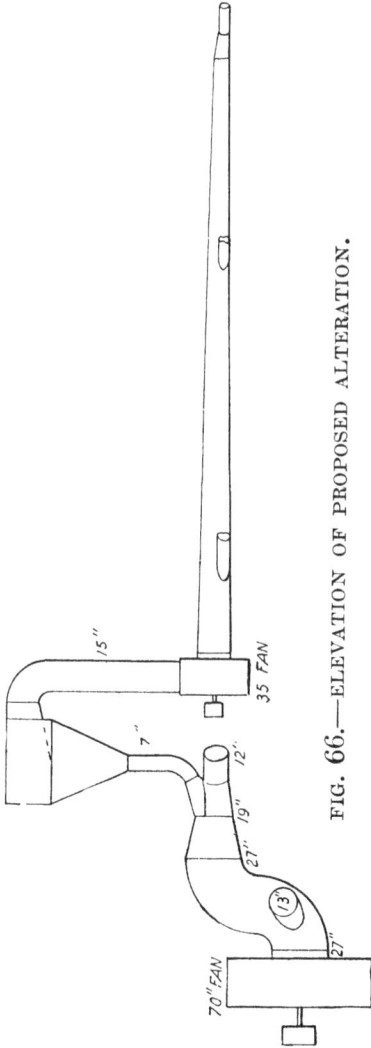

FIG. 66.—ELEVATION OF PROPOSED ALTERATION.

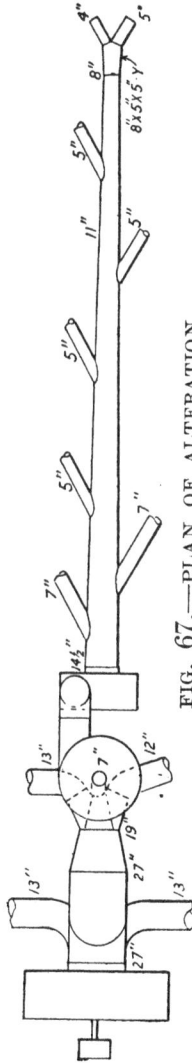

FIG. 67.—PLAN OF ALTERATION.

tions per minute. It has not enough suction about
1 inch back of elbow at fan. I would like some advice as to what change to make to increase suction.
The pipe passes over a rope transmission, as
shown. How would it work to square the pipe at
that point so as to get same area of the 27½ inch
opening of fan? I should be pleased if you can
give me some advice regarding this.

Answer.—Assuming that the three 13-inch pipes
and the 12-inch pipe are working to their full capacity—the sketch does not enlighten us on this
point—and assuming also that the small branches
are open at all times, which must be assumed when
capacities are figured, we find, by figuring the
capacity of each pipe and adding the whole, that
the 70-inch fan is too small by 128.33 square
inches, practically the area of a 13-inch pipe. Thus
the area of one 13-inch pipe being 132.73 square
inches, there are three 13-inch pipes shown and
132.73 × 3 = 398.19 square inches. The area of a
12-inch pipe is 113.1; we have one 12-inch pipe.
The area of a 7-inch pipe is 38.48 square inches
and there are two 7-inch pipes, 38.48 × 2 = 76.96
square inches. The area of a 5-inch pipe is 19.63
square inches, and we have four 5-inch pipes. 19.63
× 4 = 78.52 square inches. We have yet one 4-inch
pipe, the area of which is 12.56 square inches.
Now, add the areas; thus 398.19 + 113.1 + 76.96 +
78.52 + 12.56 = 680.33.

Now, the actual area of an A, B, C single 70-inch
fan is 551 square inches. Other centrifugal exhausters are approximately the same, 680.33 —

551 = 128.33 square inches of overload, or as before stated practically the area of a 13-inch pipe. This, of course, is figuring on the basis of a single 70-inch fan inlet being 551 inches in area. If we figure with the area of a 27½-inch pipe as a basis, the overload is much smaller, but it is the fan inlet that should be figured from in a case of this kind, where every inch counts.

However, it might be possible to load everything except the four 5-inch pipes into the main and squeeze through if the rest of the job is well proportioned and constructed, but we do not know this to be the case, because the sketch does not show what the three 13-inch pipes are doing, or, in other words, what the size of their branches is.

Now, in the absence of more definite information let us assume that this job as constructed at present will not work, and it is up to the contractor to tear it out or make it good.

The following would be the procedure: First tap the two 13-inch pipes into the 27-inch main at the fan, then reduce from 27 inches to 19 inches and make a Y-branch 19 inches to 13 inches and 12 inches and tap a 7-inch tee into its throat, or if you know the trick, make a three-way, 19-inch to 12-inches × 13 inches × 7 inches. Buy a single 35-inch fan and install it in line with the other, as shown in Figs. 66 and 67 of the illustrations, and proportion the pipe as shown.

Build a separator with the capacity of a 15-inch pipe and install it as near the big fan as possible. Make the shavings outlet 7 inches in diameter and

connect separator to 35-inch fan on the discharge
end to the 70-inch fan on the suction end.

W. H. H.

DISCUSSION ON RELIEF SEPARATOR.

From B. F. Co., Buffalo.—We are interested in
the answer to G. R. M.'s inquiry. Have you not
overlooked the fact that the total pressure against
which the exhaust fans must operate remains the
same whether a part of the shavings are dis-
tributed into the relief separator or not, since the
total length of the distributing piping remains the
same, and the velocity of the air cannot be reduced
without clogging up the system?

The arrangement shown in the sketch provides
for a relief separator, by means of which a portion
of the air handled by the fan is permitted to escape, and it is perfectly true that by this means the
size of the long run of piping which leads to the
boiler house is reduced, thereby saving in material. As each fan must operate at a pressure
which would be sufficient to convey this material
to the boiler house there is no economy of power.

Answer.—Replying to the criticism of our correspondents, it may be possible that they have
overlooked the matter of friction, which, of course,
must be considered. The precedent of making elbows with a long radius to avoid friction is based
upon data to be found in catalogs sent out by
blower manufacturers. One of these statements

is that "a short radius elbow will cause more friction than ten times its length of straight pipe," etc. Assuming that the velocity of air is the same, the loss is therefore greater in a 31-inch pipe than it is in a 22-inch pipe of the same length. In that event, an increase in horsepower, however small, is required to compensate for the loss. Let us suppose it were possible in the present case to cause the bulk of the material to flow through a 14-inch pipe. We would then have a 28-inch pipe on the relief end. Again, suppose this 28-inch pipe were only 10 feet long, as shown in Fig. 68. Now, the 14-inch pipe is running to the boiler room, a distance of 300 feet. The system would then not re-

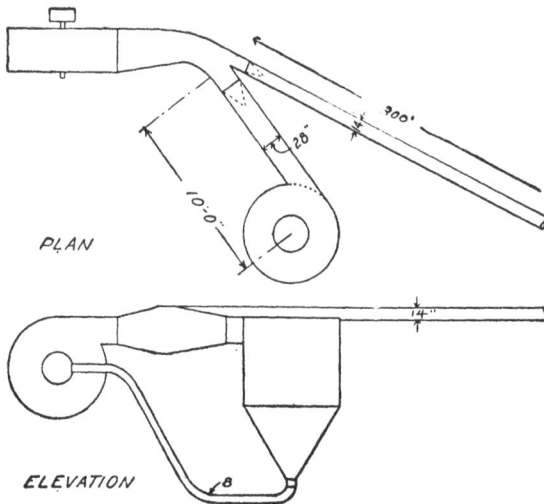

FIG. 68—PIPING METHOD UNDER DISCUSSION.

quire the same amount of power to operate it as it would if 300 feet of 31-inch pipe were used.

Our correspondent remarks: "As each fan must operate under a pressure which would be sufficient to carry this material to the boiler house, there is no economy in power." Then surely there would be no economy in power if one side of each double 50-inch fan on the present job discharged into a separator as close to the outlet as it could be placed. Following this line of reasoning to its logical conclusion there is no difference in the amount of power required to drive an exhaust fan, whether its discharge is 10 feet or 1,000 feet long.

W. H. H.

DUST COLLECTOR FOR FEATHER PLANT.

From S. S., Newark, N. J.—I have installed a dust collector in a feather plant, and some of the fine feathers escape through the air outlet of the collector. I have put a fine netting near the outlet, but the small feathers gather on the netting in large quantities so that there is no more room for air outlet which forms a back pressure in the machinery. Will you kindly advise me in what manner such collector could be built so as to do this work satisfactorily?

Answer.—In response to the query of S. S. we are enabled only to offer the correspondent a suggestion in connection with this peculiar class of work. Small downy feathers are very light, almost as much so as air, beside which they are bulky, and it is doubtful if a cyclone separator will separate them. An experiment which we recommend is to lengthen the shell A in sketch to twice the height of the inlet B, extending tubu-

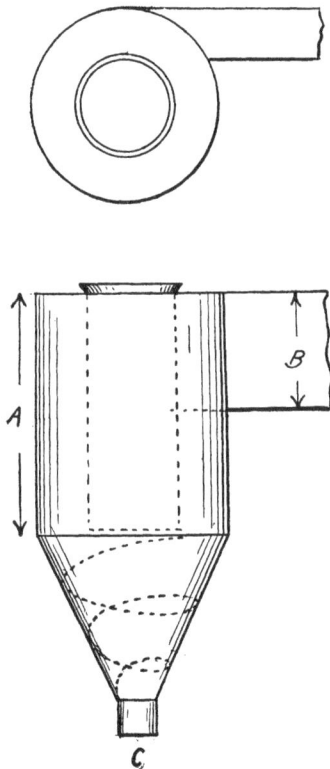

FIG. 69.—FEATHER PLANT COLLECTOR.

lar guard to top of cone and putting up a spiral 2 inches wide twice around cone ending at collar "C." This spiral must be made perfectly smooth.

ADDING RELIEF SEPARATOR TO EXHAUST SYSTEM.

From G. R. M., Philadelphia.—I have been
called upon to reconstruct the discharge end of a

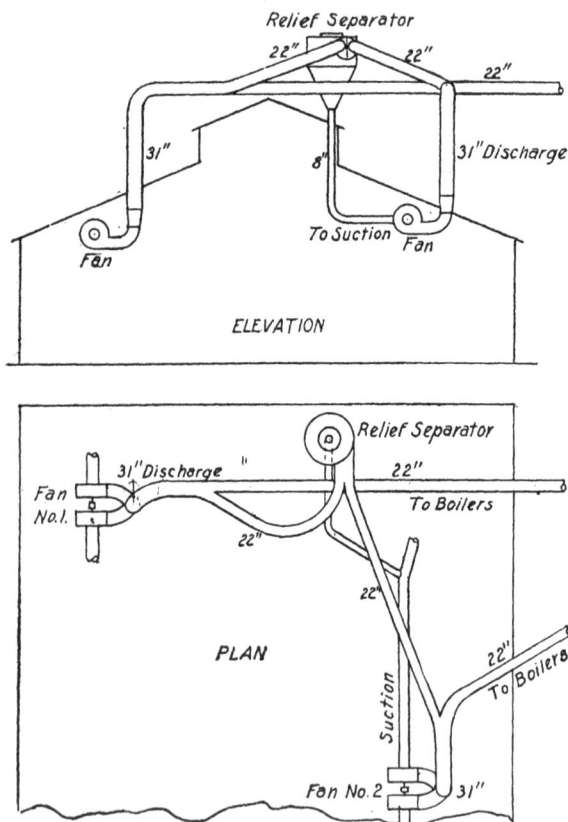

FIG. 70.—LOCATION OF FANS.

large blowpipe job in a wagon factory. The boiler
house and dry kilns were burned recently and part

of the factory, and the old discharge pipes are
burned and blistered so that they cannot be used.
The fans are located as shown in the appended
sketches, and each fan is a double 50-inch A. B. C.
Each of these fans has a 31-inch discharge pipe, yet
the discharge pipes that ran across the dry kiln and
into the separators over the boilers, a distance of
three hundred feet, were each 22 inches in diameter,
one-half the capacity of the fan. As I stated, the
old work is burned up and destroyed and I cannot find out from any one here how this reduction
was accomplished. Will you please enlighten me?
As I figure it, the two 31-inch pipes are equal in
area to a 44-inch pipe.

Answer.—In answer to query of G. R. M., I
will say that this was accomplished by what is
known as a relief system, the object being not only
to save material but to reduce the power required
to operate the system. An enormous amount of
friction is avoided here by reducing the size of
the pipes one-half, and consequently a great
amount of power is saved. The appended sketch
shows how this is done. A separator is placed in
a convenient spot on the roof of the mill, the dis-
charges of the double fans are first merged into
one, and run out through the roof; then they are
caused to curve so that the material is thrown to
one side of the pipe, elbows being used for this
purpose, as shown at fans No. 1 and No. 2.

Then Y branches with one leg straight are connected as shown, care being taken that the straight
leg is placed so that the bulk of the shavings are

forced into it; in other words, it must be placed on the heel side of the elbow. These Y branches in this case are each 22 inches × 22 inches × 31 inches, thus dividing the pipe into two 22-inch pipes again.

The outside leg of each Y branch is then connected to the boiler house, the inside legs to the relief separator. It will be seen that this arrangement not only assures an equal division of air all around, but that the quantity of shavings blown into the relief separator will be comparatively small; in fact, must be, or it will be too much for the system to handle over again.

The relief separator material outlet must be piped into some convenient place; in either of the many sections care must be taken to insure capacity wherever it is tapped in.   W. H. H.

REMEDY FOR DUST PIPE CLOGGING.

From R. C., Indiana.—About a year ago I connected a double buffing wheel with an exhaust system. The exhaust from the wheel was connected to a large chimney and worked satisfactorily at first, but soon there was trouble with it stopping up continuously. The dust from this wheel had to travel about 25 feet, and through two 45-degree elbows before reaching the chimney.

I used 6-inch pipe in the first attempt. I tried to remedy the trouble by putting a box 18 inches wide, 18 inches deep, and 36 inches long in the line about 6 feet from the fan, with the pipe tapped into the ends of the box. I thought this box would act as a sort of settling chamber, that the dirt would drop to the bottom of it and the air would go on up the chimney. I found, however, that enough dirt would carry across the box to stop up the pipe.

This buffing wheel is in a cellar which has an 8-foot ceiling and is about 90 feet from the nearest point in a wall through which I could allow it to exhaust into the open air. It is used for polishing the brass fronts on cookie cans after they have been washed and have gone through the drier. To move the wheel to any other location would mean unnecessary handling of the cans.

The dirt does not amount to more than 2 bu. a week, and the chimney to which the exhaust pipe is connected has some exhaust steam in it. There is no other chimney within 50 feet to which this exhaust pipe could be connected. The building is six stories high, and the space on the different floors is so utilized that a pipe cannot be run to the roof.

I will certainly appreciate any help which readers can give me on this problem.

Answer.—(From American Blower Co.)—Referring to the trouble which R. C. is having with his exhaust system, as outlined, we believe that this could be very nicely taken care of by the installation of one of the smaller sizes of Morse Rarified Dust Collectors. It is possible that, due to the very fine quality of the dust, all of it would not be taken out of the air. Enough of it, however, would be taken out so as to eliminate the trouble which occurs in having the chimney pipe stopped up.

Much better results could be obtained by using a spray in connection with the separator. In this manner practically all of the dust would be taken out of the air. This, of course, would be more expensive since it would mean piping for the water supply and also a drain connection to a sewer, and would also include the cost of a spray nozzle.

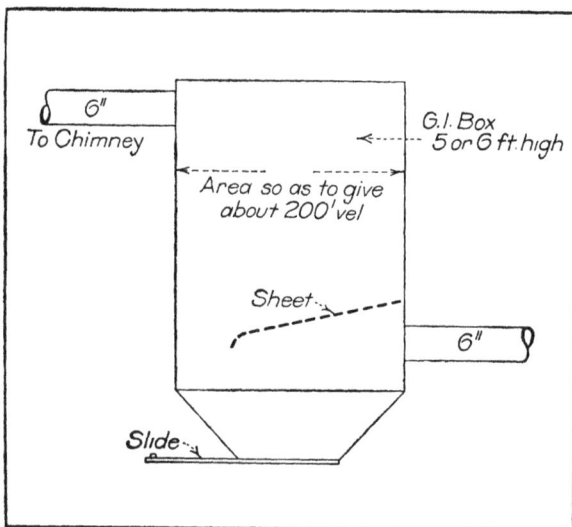

FIG. 71.—REMEDY FOR DUST PIPE CLOGGING.

In regard to using the 18 by 18-inch box, in our opinion the trouble is that the area of this box is too small for the volume of air which is being handled. Good results should be obtained from this method, providing the area of the box is sufficiently large so that the velocity of the air passing through it would drop to about 200 feet per minute. This low velocity should allow the dust particles to be deposited in the bottom of the box.

A box of this kind with an area to give a velocity of 200 feet per minute should be about 5 or 6 feet high, with the pipe entering near the bottom. The outlet pipe should then be taken off near the top. It would be well to put in a deflector sheet, horizontally, just above the entering pipe, for about two-thirds the distance across the box. We believe that this arrangement would catch most of the heavy particles.

The easiest and cheapest way to overcome this trouble would be to run a 6-inch duct over the distance of 90 ft. and let it discharge into the atmosphere. We do not consider 90 feet as being a very long run of pipe. If necessary, a small box could be placed over the end of the pipe so as to prevent the dust particles from being scattered around in the air.

CHAPTER XV.

Hints on Installing an Exhaust System.

By "M. F. H."

Assuming that you have received the contract, it is now in order to execute it with satisfaction to your customer and profit to yourself. You may remember that you made a list of the material required when you were making up your price. This list should have comprised everything required on the job, each piece under an appropriate heading. You may now hand an order to your foreman, which will be a copy of your list, or it can be divided, the pipe being turned over to the apprentices, the elbows, collars and reducers being given out in a lump quantity. The hoods and special fittings should be given to the best man in the shop, who should be assisted by carefully prepared drawings. Many of the hoods will have reinforcing bars or angles, and it is well to get them out, punched but not riveted, to the piece, as quite a good deal of cutting and fitting is necessary at the machine.

It is of the utmost importance that you should have correct data from which to estimate the cost of the work, not only as the job proceeds but for future reference. Any good system of cost sheets will secure this, but as the writer has been employed in several shops where the cost of a job was more or less of a guess, he has been obliged to devise a method of keeping in touch with the cost of work which has stood the test of many years' operation. Supply your men with a few pads of charge slips similar to the one herewith reproduced, or a modification of it. Have each man state under the heading "Nature of Work," whether they were making or erecting. This slip will tell you all that you ought to know and does not involve much writing. One slip should be used for every order given to the workman, and if a box is provided where they can be deposited you will have every morning a complete record of the preceding day's work. It is an advantage to have each man weigh his own work, as you then have the weight and the time on one slip. As soon as the list is turned into the workshop an account should be begun in a contract book wherein should be entered every item made or purchased for that particular job. If you purchase anything, such as lumber for braces, scaffolds or platforms, you should mark the quantity and the price on a slip and turn it in along with the others. The habit of putting everything on a slip will soon become so well fixed that there will be little chance of anything being overlooked. It is only the work of a few minutes to copy the data from the slip into your book, and by a little adding you can tell at any time just how you stand.

The time of the men for the payroll can be made up from the same slips by copying the time into another book, together with the name of the customer. This enables you at any time in the future to tell which of your men worked on any certain job. As soon as the shop work is finished the account in the contract book may be added up and you will then know whether this part of the job has been made up at a profit or not. If it comes out as you had figured that it would then you have done your work well; if not, look for the reason, and, when it is found, do not merely promise to yourself that you will be more careful next time, but make a note stating the reason and insert it as part of the permanent record of the job. It is better to have it perpetuated in your contract book than in your bank book.

Take your men and material now to the building and begin to erect. If the fan has been delivered it is well to have it put up at once, and it is an advantage to support it on a platform hung by means of pipe hangers from the ceiling. Four standards of three-quarters or 1-inch pipe should be cut and threaded, making the threads rather long. The lower ends are passed through the platform supports and the floor plates screwed on, bearing on the under side of the timber. A few turns on each plate will level the fan, and when this has been done each plate should be screwed to the platform to prevent any shifting. It is best to start two gangs simultaneously, one on the inlet side, the other running the discharge pipe to the separator. The inlet mains with the

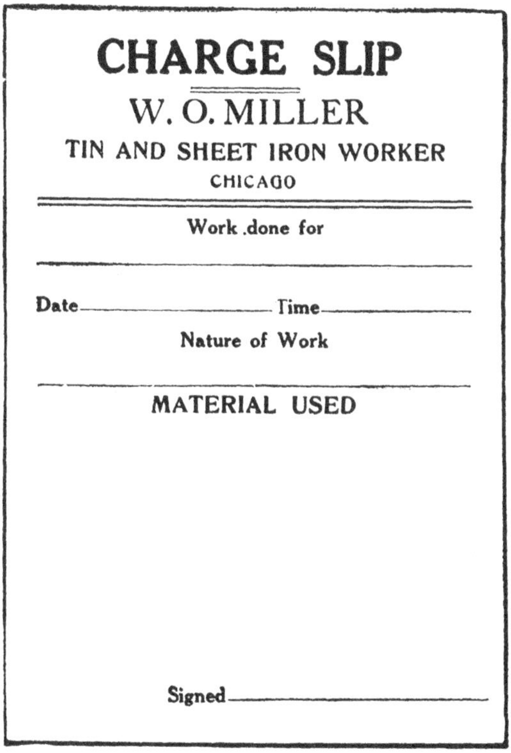

CHARGE SLIP
W. O. MILLER
TIN AND SHEET IRON WORKER
CHICAGO

Work .done for

Date——————— Time——

Nature of Work

MATERIAL USED

Signed——————

FIG. 72.— FORM OF CHARGE SLIP.

branch collars can be hung in place before any of the machines are connected.

Of course you took care to send in your bill for a third of your contract price if your terms were made that way. As the work proceeds your system of cost-keeping will enable you to tell from day to day whether you are ahead or behind in the game, and as you may be asked to do little things that were not definitely included in your contract you know just how much of that sort of thing you can afford to do. If your men are "pulling out" well, it pays to spend a while touching things up here and there, and by all means, if you can afford to do it, give your customer just a little more than he bargained for, tactfully seeing to it that the fact does not escape his notice.

WHEN TO LOOK FOR TROUBLE.

If trouble is coming your way it is scheduled to arrive about the time that the fan is started, at least no one looks for it till then. The gentleman who feeds the planer and knows your business better than you do, while "swapping plugs" with his neighbor, has confided to him the information that "that hood will never catch the shavings."

The presiding genius of the "sticker" never saw a machine piped up like that and he sketches weird and impossible things on a board, illustrating how he would do it.

This is all a part of the game that you have started to play, and, if you look at it in the right way, those men will become your best educators. They are the human factors that you must learn to take into account when making up your price in the future. You cannot ignore them for, after all, it is they who have to use the apparatus, and the more intelligent of them can give you many a useful hint. If the "prophets have erred" you can shake hands with yourself; if not, let us look for the trouble.

You know, of course, that your fan size and speed are "according to Hoyle" (or Hayes), yet here is a machine that runs only a few minutes before the cutters are choked with shavings. There may be one of several reasons for this; the material as it leaves the knives may strike a part of the hood that throws it back on to the head instead of deflecting it up into the pipe. Note the angle made by the shavings or chips and the back of the hood, for they will leave the hood at the same angle, increased a little by the pull of the fan. If this point is all right, next see to it that there is sufficient air opening at the mouth of the hood to satisfy the demand of the fan, and to keep up the volume at the necessary velocity.

Try to arrange your hoods and pipes so that the material will travel on to the fan with no abrupt change of direction, and sometimes, as in the case of side moulder hoods, you may have to put in deflection strips to do this. Suppose that your main or branch pipes gradually fill first with large chips and then with the finer shavings. You are either not giving the pipe sufficient air at the hood or your fan is not traveling fast enough. The system is then like a river whose velocity or volume is not sufficient to carry along the stones and sand that it holds in suspension, and it deposits them along its bed.

To know exactly what the fan is doing you can use a gauge that is made as follows: Take two pieces of quarter-inch glass tubing and unite them by slipping a piece of rubber tubing over the ends, the combined length should be about fifteen inches Bend the rubber tube into the form of the letter U and pour sufficient water into the gauge thus made to fill the glass tubes up to within four inches of the top. Slip a rubber band over each glass tube to act as a register and using a second rubber tube connect one leg of the gauge with a quarter-inch hole punched in the galvanized iron pipe. The water will be raised in one leg and lowered a corresponding distance in the other, both movements being followed by slipping the bands, one up and the other down. Measure the difference with your rule and note the amount when the suction is good. There should be very little change between the displacement close to the fan and that obtained at the more remote parts of the system if the pipe proportions are correct.

The use of this inexpensive little instrument will soon enable you to tell whether or not you are getting the air movement necessary to do the usual work in a shaving exhausting system. Sometimes the indiscriminate use of the blast gates will cut down the air volume to such an extent that the larger pipes will fill even after the job has been accepted, but a little explanation made to the foreman generally puts a stop to this trouble. It frequently happens that satisfactory results are obtained during the first few days that the system has been running, and then complaints begin to come in—"suction is getting weak," "pipes are filling up," etc. The trouble nine times out of ten is caused by a belt stretching on either the countershaft or the fan.

You may find that the shavings will keep circulating around in the separator until they gather in a heap and choke the tip. The remedy for this is to rivet a narrow band not over an inch and a half wide spirally around the inside of the lower part of the cone, extending upward about two feet. This band also helps to send the shavings with a greater velocity into a furnace feeder. As the heart of the whole system is the fan, always provide means for readily examining it, for a block may break or bend the blades, and this is a fact that should be kept in mind in case of trouble.

Go to all reasonable lengths to make your job a success before any complaint is made, and let every difficulty be welcomed, overcome, and thereby made a stepping-stone toward future success. You have started a line of work that calls for a high degree of intelligence and adaptability, not everyone can do it successfully, and nothing can carry you through without painstaking endeavor.

It is a good plan to label the most conspicuous hoods with your name and address, and the job will then stand as an advertisement of your ability.

CHAPTER XVI.

HINTS ON ESTIMATING THE COST OF AN EXHAUST
SYSTEM.

By "M. F. H."

The recent admirable articles on "Practical Exhaust and Blow Piping," by William H. Hayes, in your magazine, have no doubt given rise to a desire in the minds of many sheet metal workers to enter this profitable and enlarging field. Several reasons have in the past deterred many from going after this class of work, the principal one probably being a lack of confidence natural to the inexperienced. It is with a view of helping such that the following hints are being penned, and they are all based upon the actual experience of one who has been "through the mill" and who has spent not a few dollars learning how not to do it.

I propose to go with the novice from the moment he receives the inquiry to the point when he leaves the mill office with his well-earned check in his pocket, and intend that the whole transaction will have that invaluable characteristic of all successful business deals, viz., that both buyer and seller will gain.

Let us suppose that you have received a short letter from Messrs. Logwood & Co., stating that they are in the market for a shaving and dust exhausting system, indicating their wish to have your figure. Reply immediately, stating that you will call at a certain time, and at that time be there; not in a half-hearted, hesitating sort of way, as if you were not quite sure about it, but bright, alert, and self-confident, because your every word or move will indicate to the man who is to give you a nice contract whether you are the proper man to successfully carry it through or not. See the foreman of the mill as soon as possible and find out which of the machines he wants to have "connected," noting in the case of those having more than one head whether or not the whole machine is to be piped. It is well now to make a rough plan of the mill, which we will suppose has only one story. Mark on this plan the location of the machines, spacing them correctly by measurements from the end and side walls. A small square can be used to denote a machine, but it should have a circle in the center for a top feed machine such as a planer, or at the side in case of a jointer. This circle should show upon which side of the machine the riser can be placed. and within the circle may be a figure indicating the diameter of the pipe, according to the following table:

```
24-in. single surface...........................6 in.
30-in. single surface...........................7 in.
24-in. double surface.............6-in. bottom and top
30-in. double surface........6-in. bottom and 7-in. top
Moulder or sticker......6-in. top, 5-in. sides and bottom
Tenoners.....................5-in. top and bottom
Upright molder .............................5 in.
30-in. Invincible sander........8 in., belt sander 6 in.
16-in. jointer ..............................5 in.
18-in. jointer ..............................6 in.
Cut-off saw .................................4 in.
Rip saw ....................................5 in.
Band saw...................................4½ in.
Swing saw..................................4 in.
```

Consult the foreman regarding the location of
the floor sweeps, pointing out the fact that all of
the under feed machines will connect to a tee,
which can be used as a sweep by opening the
cleaning cap. Don't worry at this stage about
the location of the separator, the fan or the shape
of the hoods; take your difficulties one at a time,
systematically. Get your layout right, that is
your business for the present. With the data
that you now have you are in a position to determine the size of the fan by adding together the
areas of the branch pipes. If the mill is long and
narrow you may find it best to use a double fan,
in which case you divide your total area by two.
If the mill is nearly square then a single fan can
be used, and it should be placed nearest the machines that do the heaviest work. Settle in your
own mind where the fan is to go, and then consult
the foreman as to how he is to drive it. You will
probably be asked at this point to state the speed

that you will require and the diameter of the pulley on the fan, all of which you should get from the catalogs of the fan makers. Do not accept his statement about the speed of the line shaft. Nine times out of ten it is a guess, and you cannot afford to take any chances. Put an indicator on the shaft yourself. Multiply the diameter of the fan pulley by the number of revolutions that you desire and divide the product by the number of revolutions of the shaft. The quotient will be the size of the driving pulley. Care should be taken to see that it can be used without cutting into joists. Avoid the use of a crossed belt, especially if the drive is short, and insist upon a belt the full width of the face of the fan pulley. If, after considering everything, you find that the fan can be located in the place first selected, your next move is toward the separator, which most frequently is placed near the boiler house.

Tell the mill foreman just how you propose to run the discharge pipe from the fan so that it may not be in his way, for it pays to get his approval before anything is made.

You should now run lines on your drawing which should show the separator and fan, indicating the dimensions of the mains and branches, noting all that comes in the way. Avoid offsets as much as possible, but when they are unavoidable put a cleanout door close to them in case the pipe chokes at this point. It is well now to

go over the machines one at a time and make
sketches of the hoods or hoppers. Note the direction in which the material is thrown from the
cutter, and so shape your hood that it will gather
the material toward the pipe. The swiftly revolving cutter is itself a fan, and your hoods must be
strong enough to withstand the impact of blocks
and chips. Some machines, such as moulders and
tenoners, have movable cutters, and various de-
vices are used to take care of them, the common-
est of them being the cast iron bale joint.

You will be greatly benefited by a walk through
a mill already equipped with a shaving exhausting system, and you should seek every opportunity
of getting "wrinkles" in this way. If this is
impossible and your own ingenuity is not equal
to the occasion, then it would be better to seek
advice through the columns of SHEET METAL
regarding any difficulty that you may have
in designing the various fittings required. It is
well to familiarize yourself with the operation of
the various machines, so that your fitting may be
adapted to the wide range of cuts that some machines can make.

You have now all of the data that you require
in order to make up your estimate, and your
business at the mill for the time being is at an
end. Your office work begins by making a scale
drawing, showing the location of the fan, pipes
and machines. To "take off" the quantities, it
is well to begin either at the fan or the machine

farthest from the fan and work toward the other end. Note on your list all of the items as you go along, or have a headed list and place each price under its appropriate heading. Either way will give satisfactory results, provided nothing is omitted. Do not let prices enter your thought while you are working on quantities. Try to picture each machine in your mind complete, so that such items as blast gates, telescopes, lug bands, swivels, etc., will not be overlooked. Standard fittings, such as elbows, tee joints, collars and pipes can be priced from your regular list, but experience only can guide you in estimating the cost of hoods, hoppers, etc. No rule can be given, but after a few jobs have been completed an average cost may be obtained that will enable you to estimate by the pound. In the meantime it is safest to use your best judgment as to the value of each piece. The same is true regarding the time required to erect the work. Go over the plan and get an idea of the time that you think will be taken on each machine, then on the risers and main pipes. Allow a fair margin for interruptions, changes and all of the little "snags" that you will surely encounter. Take the capacity of your men into account and do not base your calculations upon how long it would take you to do the work personally.

Assuming that you have settled upon a price, write out your proposition, stating clearly what you propose to do, the number of machines that

you intend to connect, the gauges of metal that
you will use, adding perhaps the usual little joke
about "first class manner" and "best workman-
ship." Beware of the much-abused word "guarantee," unless you can give it a meaning not
liable to misunderstanding. For instance, you
carelessly guarantee to remove all of the shavings
from the machines, which you find later on that
you cannot do, but the wording of your letter
or contract form has made the impression on
the mind of the purchaser that his mill floor will
be as clean as his office floor, and he may not be
satisfied with results as good as can be obtained
in practice. Your motto should be "Promise little
and do much," assuring your customer that the
results will be satisfactory. It is a good plan to
make your terms of payment one-third on de-
livery, one-third on completion, and the balance
in thirty days. This arrangement is fair to your
customer and yourself, and somehow he takes
more interest in something that he has paid his
money for than he does if you pile in your work,
which to him has no present value.

You may in time be asked to pipe up a machine
or apply a fan to an entirely new purpose, and it
is more satisfactory all round, if there is any
doubt as to the result, to promise merely to do the
best that you can. It is unreasonable to expect
you to experiment at your own expense and risk
when some one else is to derive by far the greatest
benefit. Unless the job promises a fair degree

of success it is better to let it go, unless you can get a price that will pay you for the time that you may have to spend upon it in order to make it work satisfactorily. It is advisable to put your sketches, cost sheet, etc., together with all of the data pertaining to the job, into a large envelope and take it and the plan with you when you go to "talk up" the installation. The points to be emphasized in your talk will depend largely upon your customer; it may be price or quality, and generally it is both, but remember one thing, your work will be a credit or reproach to you long after the price paid for it will be forgotten. A very large majority of the men who will have to use the system will in a few years be working in other shops, and it depends upon you whether they will be "boosters" or "knockers" for you when sheet metal work is under discussion

CHAPTER XVII.

DETAILS OF SETTLING CHAMBERS FOR GRINDERS.

HOW TO DESIGN AND CONSTRUCT THIS IMPORTANT PART OF AN EXHAUST SYSTEM—THE MODERN WAY OF PREVENTING IT FROM CLOGGING.

BY "SEATTLE."

In many exhaust pipe systems for conveying away the dust from emery wheels the pipe is carried direct from the hood over the wheel to the main exhaust pipe. As a rule there is an elbow in the piping near the hood, and the system is likely to clog frequently at this elbow if some means are not provided to forestall just such an occurrence. It is therefore customary to place a clean-out at this point. Then, when the clogging does occur, it is an easy matter to poke the solid matter out through this clean-out and thus keep the system free.

In a large plant built for war purposes the modern method of providing settling chambers near the grinders was specified by the designing engineers of the exhaust system. There were batteries of grinders, practically all of which were

twin machines, that is, two grinder wheels mounted
on a single shaft.

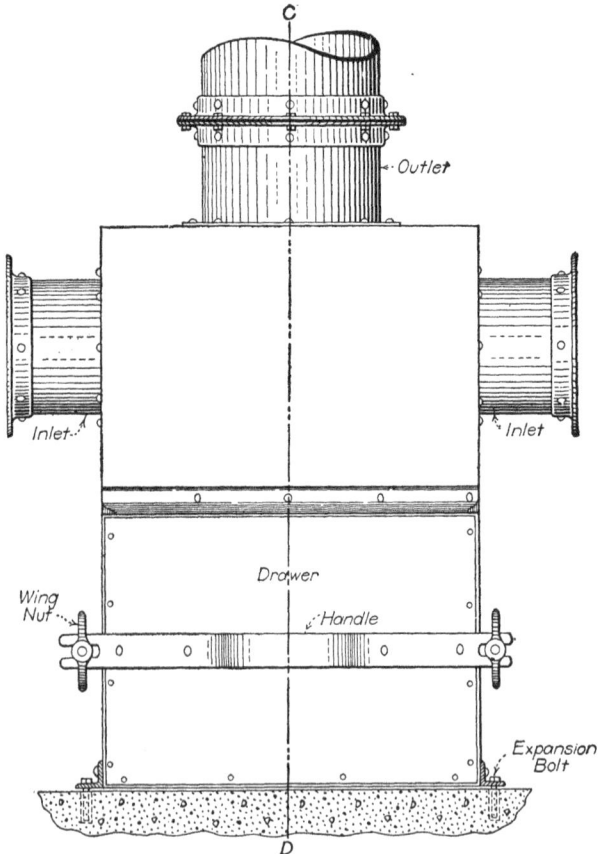

FIG. 73.—FRONT ELEVATION.

Directly back of each machine, and securely
fastened to the floor, was placed one chamber, so

that for these machines there were two inlets and one outlet, as can be seen in Fig. 73 of the accompanying line drawings.

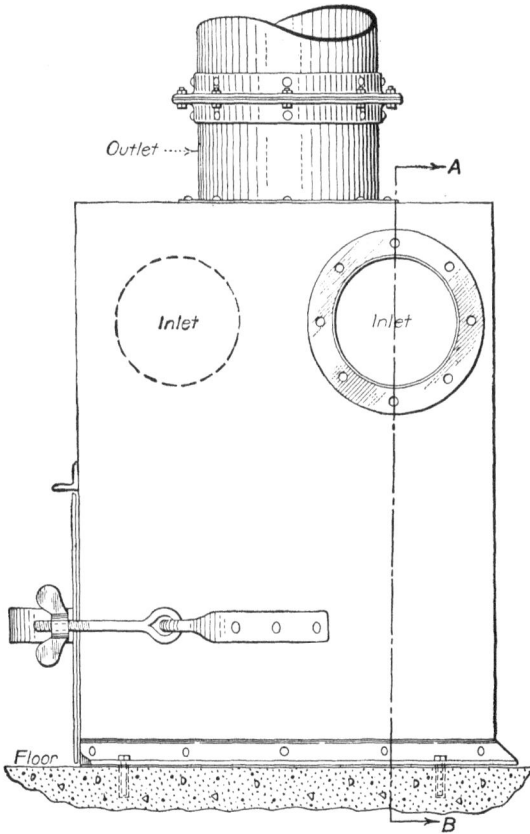

Outlet ····>

A

Inlet

Inlet

Floor

B

FIG. 74.—SIDE ELEVATION.

These chambers are simply nothing more than boxes made of fairly heavy sheet metal—20-gauge

galvanized sheet steel—and amply stiffened by
angle irons. They were about 2 feet high, 18 inches

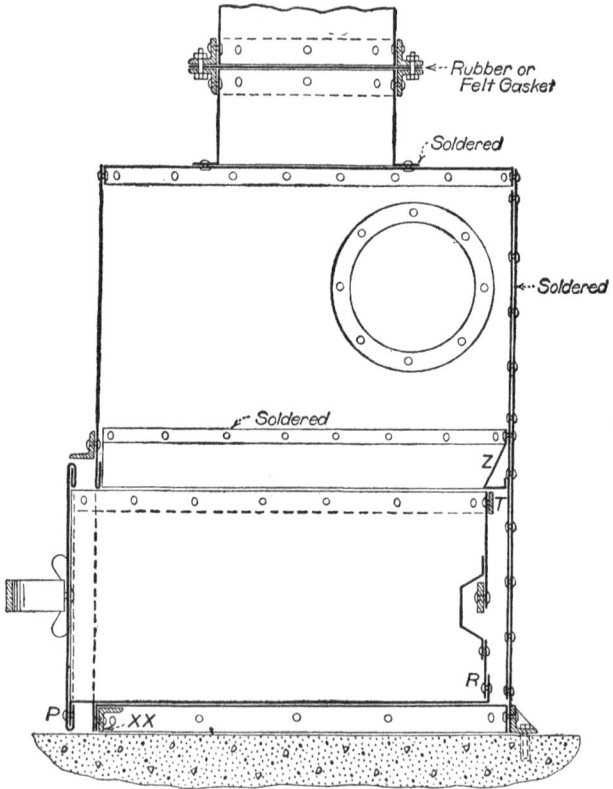

FIG. 75.—VERTICAL SECTION ON CENTER LINE C D,
FIG. 73.

from front to back and 15 inches wide. The
drawer is about 10 inches deep.

Actual measurements may vary from these dimensions slightly, as an attempt was made to cut the material from the stock available with the least waste possible, and a variation of 1 inch or so either way was permitted by the engineer to do

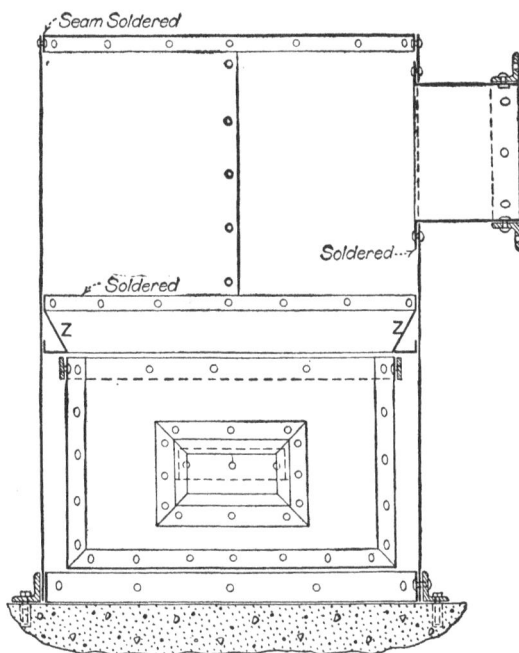

FIG. 76.—VERTICAL SECTION ON LINE A B, FIG. 74.

this. The inlets and the outlets, however, had to be to the size specified as follows: 5 inches internal diameter for the inlets and 7 inches internal diameter for the outlet.

As may be seen by the drawings, the inlets are
tapped into the sides and off centers, that is, they
do not directly face each other. If they were
placed directly facing each other the pull of the
exhaust fan might be minimized.

The outlet is tapped into the top of the chamber,
to which is connected a vertical pipe, leading to
the main exhaust, which it enters at an angle horizontally. The turn from vertical to horizontal is
negotiated by a long radius elbow. To the inlets
are naturally connected the piping and elbows
from the hoods over the grinding wheels.

A vertical section on the center line C D is shown
in Fig. 75. In this sketch the drawer is practically
pulled out. In Fig. 76 is shown a vertical section
on line A B of Fig. 74.

The body of the chamber was made exactly like
a section of square pipe, that is, square bends were
made at the corners, the seam coming in the center
of the back as shown in Figs. 75 and 76.

The isometric sketch in Fig. 77 perhaps gives a
better idea of the way the body was made. Note
how the opening for the drawer is cut out of the
body with the stop edge A A provided at the corners; also how the stiffening hem edge is turned in
on three sides of the opening, and again note that
this hem edge is notched away at the bottom so
that there is always the same amount of metal all
around the opening to insure a true contact surface
for the stop edges of the drawer. This is important, as it is absolutely essential to prevent air
leakage.

The top for the body is just a pan; likewise the bottom; that is, these pans are merely flat sheets with the proper riveting edges turned up square as shown in Fig. 78. The top pan is not notched at A A of Fig. 78, and was riveted to the body as shown in Figs. 75 and 76, while the bottom pan was riveted to the body in just the opposite way, also as shown in Figs. 75 and 76.

The notches at A and A of Fig. 78 would coincide with those on the edges A A of Fig. 77, and are provided for the same reason as mentioned in connection with the edges A A of Fig. 77.

It is to be understood that all rivet holes, holes for the inlet and outlet and for other purposes were located and punched on the sheets before bending into shape.

The inlet and outlet collars are naturally only short pieces of pipe of the required diameters. They have a suitable flange stretched off for riveting the collar to the body.

The outlet flange was placed on the outside of the body as shown in Figs. 75 and 76, whereas the flanges of the inlets were placed on the inside of the body as shown in Fig. 76. This was done so as to have all edges with the direction of the flow of air, thus giving a smoother surface, and consequently less frictional resistance to the fan.

A 1 by ⅛-inch angle iron was curled to fit around the edge of the collars and riveted thereto as shown in the drawings. To these angle irons were bolted the angle iron on the connecting pipe, with a suitable gasket between.

The stop edges A A of Fig. 77 are stiff enough for the purpose, but the edge B if not reinforced would undoubtedly sag; hence the stiffening 1 by ⅛-inch angle iron shown in the drawings. At the bottom of the body and at the sides and back the same size angle iron was riveted on as shown. The rivets holding these angles are the same ones which hold in the bottom pan, as can be seen.

The same remarks apply to the front angle X X of Fig. 75 except that this angle is placed behind the pan, and only the two end rivets are attached to the body because the drawer opening comes at this point.

The ends of the exposed angles of the sides at the front, as well as the stiffening angle above the drawer, are rounded off for a finish as shown, while at the back, instead of mitering the side angles with the back angle, the back angle is cut off square and flush with the side of the body, while the side angles extend as much as the width on the floor of the back angle, the ends of the side angle being cut off on a slant as shown in Fig. 74 for a neat finish.

The drawer was made small enough to give plenty of clearance with the opening of the body, allowing also for the 1 by 3/16-inch band iron T of Fig. 75, which stiffens the three sides of the drawer. Inasmuch as this drawer is considerably smaller than the body of the chamber at the sides and at the back, and as most of the sediment falling from the inlets would travel along the sides of the body, it follows that much of the sediment would land between the body and the drawer. To obviate this, shedding strips bent to the shape shown at Z in Figs. 75 and 76 were riveted to the sides and back of the body.

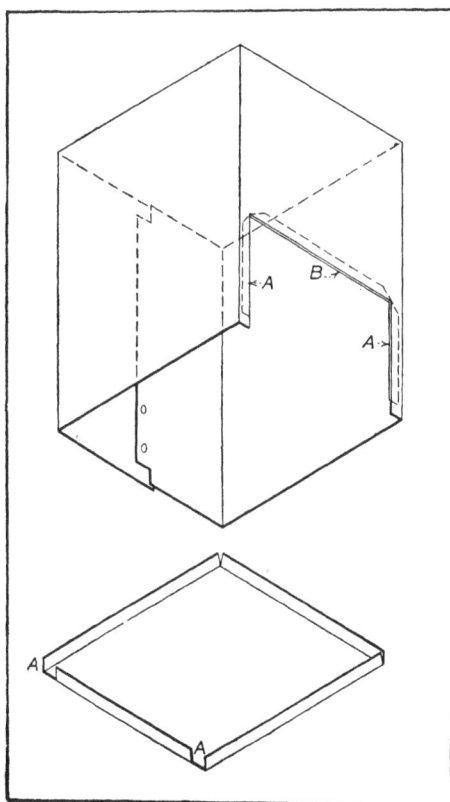

FIG. 77.—(ABOVE) ISOMETRIC DRAWINGS SHOWING HOW BODY OF CHAMBER WAS MADE.

FIG. 78.—(BELOW) BOTTOM AND TOP OF BODY.

The two sides and the bottom of the drawer are one piece bent up U shape and a ⅝-inch edge was turned out square all around at the front, to which the front piece of the drawer was clinched and riveted, as shown at P in Fig. 75. At the back of the drawer at ¾-inch edge, in which the rivet holes had previously been punched, was turned square inward all around, to which was riveted the back of the drawer as shown at R in Fig. 75.

This back piece of the drawer is only a flat sheet, which has a slit cut in it about central, 76 inches long horizontally and 2½ inches deep. Along the top edge of this slit a heavy piece of heavy band iron was riveted. This forms a strong and comfortable handle by which to lift the drawer in connection with the handle on the front of the drawer when it is removed from the chamber.

Over this slit a box was riveted as indicated in Figs. 75 and 76, which also partially shows the band iron handle on the front of the drawer. This handle on the front of the drawer was made from 1¼ by 5/16-inch band iron long enough to bend out U shape, as per Figs. 73 and 74, and to project 1 inch beyond the sides of the drawer.

The ends of this band iron were notched as shown to receive the threaded ¾-inch rod, one on each side of the chamber. This rod had an eye formed at the back end engaged into an eye formed on a piece of band iron which was securely riveted to the side of the chamber as shown in Fig. 74.

When the drawer is pushed into the chamber the rods on both sides of the chamber are brought into the notches on the band iron on the front of the drawer when a wing nut screwed up tight to the band iron forces the stop edges of the drawer tight against the stop edges of the chamber, preventing air leakage.

It is to be understood that all riveting like at P and XX of Fig. 75 was flush headed so that there is a smooth surface. All wrought iron was galvanized and the chambers were painted to conform to the color scheme of the other installations.

INDEX

www.ingramcontent.com/pod-product-compliance
Lightning Source LLC
Chambersburg PA
CBHW031359180326
41458CB00043B/6541/J